Progress in Vehicle Aerodynamics and Thermal Management

Jochen Wiedemann
Editor

Progress in Vehicle Aerodynamics and Thermal Management

11th FKFS Conference, Stuttgart, September 26–27, 2017

Springer

Editor
Jochen Wiedemann
Research Institute of Automotive
Engineering and Vehicle Engines
Stuttgart, Baden-Württemberg
Germany

ISBN 978-3-319-67821-4   ISBN 978-3-319-67822-1 (eBook)
https://doi.org/10.1007/978-3-319-67822-1

Library of Congress Control Number: 2017961794

Printed on acid-free paper

This Springer imprint is published by Springer Nature
The registered company is Springer International Publishing AG
The registered company address is: Gewerbestrasse 11, 6330 Cham, Switzerland

Welcome Message

The individual freedom given by a personal road vehicle is likely to continue. It is still an important attribute of our life. In response to the environmental challenges, the incentives to improve the efficiency of road vehicles are strong and continuously growing.

Following the recent discussions related to emissions, the electrification of cars seems to be the principle path of future development processes. Despite several specific aero-/thermal concerns for electrically driven vehicles, the overall aerodynamic issues for both electric cars and conventional ones are similar. However, it should not be forgotten that the efficiency of aerodynamic measures is enhanced in the case of electric power trains because of possible regenerative breaking.

Consequently, continuous reduction of driving energy for road vehicles remains the main task without forgetting passenger comfort, driving stability, and thus safety, internal/external noise generation, and thermal management.

The main topics of this year's conference are on new vehicles, new on-road, and new cooling/thermal management results with a special focus on unsteady aerodynamics.

Usually, the aerodynamic development process of vehicles in wind tunnels relies on steady-state conditions disregarding any unsteady aerodynamics present on the road. Low turbulence intensity flow and fixed test objects at zero degree to the wind are the main boundary conditions.

On the road, the vehicle almost never experiences these conditions, and thus, more and more attention is paid to reduce the gap between steady-state wind tunnel results and later behavior under real conditions. To diminish this gap, tools and methods are discussed nowadays to install or improve simulation techniques in wind tunnels. Achieving early information or predictions on car handling will reduce the development process considerably.

Joining the FKFS-Conference is an ideal opportunity to meet experts from industry, universities, or other institutions; to exchange new ideas; and to learn about the latest developments, methods, and tools.

Traditionally, as a unique feature, the FKFS-Conference offers live demonstrations showing measurement technologies applied in the FKFS wind tunnels and laboratories.

August 2017 Jochen Wiedemann

Contents

Thermal Management

A Coupled Simulation Approach to Race Track Brake Cooling for a GT3 Race Car

Will Hunt[1], Adam Price[1], Sacha Jelic[2(✉)], Vianney Staelens[3], and Muhammad Saif Ul-Hasnain[2]

[1] Bentley Motors Limited, Crewe, UK
[2] Exa GmbH, Landshuter Allee 8, 80637 Munich, Germany
sacha@exa.com
[3] Exa UK Limited, London, UK

Abstract. During race track operating conditions, the vehicle is constantly accelerating and braking from high to very low velocities. This generates a lot of heat that needs to be absorbed by the brakes. Sufficient cooling is required to prevent the brakes from overheating. When brakes exceed their critical temperature, they can lose grip and start fading. Brakes can lose quite some heat through radiation and conduction to their surroundings, but most of the heat can be released through cooling airflow convection. Improving the cooling airflow to the brake discs can significantly lower the brake disc temperature during the race track duty cycle. An efficient design for convective cooling will avoid large drag penalties or significant brake disc weight increase. With simulation, the brake disc and brake system design can be optimized more efficiently, to allow more cooling airflow, by visualizing the flow and it can be used in early design stages. The 3D CFD simulation method is coupled to a radiation/conduction tool to include radiation, conduction and convection effects. It can predict the brake system temperature over time during the race track duty cycle. The results have been compared against experimental data and several design variants have been tested to improve the design.

1 Introduction

This chapter presents the simulation results of a GT3 type of vehicle for a race track brake cooling scenario. The brake disc temperature during a race track lap is shown and compared with experimental data. The lattice Boltzmann based 3D CFD solver is coupled to a radiation/conduction solver, to include 3D convection, conduction and radiation effects. The wheels and tires are rotated realistically.

Brake cooling simulations have already been published in the past with a similar coupled CFD tool [1–5]. This method has now been applied in a very similar way for a race track type of duty cycle and a GT3 type of vehicle. The difference between a standard duty cycle and a race track scenario is that the duty cycle repeats the exact same braking and acceleration cycle many times. It applies the same acceleration and

J. Wiedemann (ed.), *Progress in Vehicle Aerodynamics and Thermal Management*,
https://doi.org/10.1007/978-3-319-67822-1_1

deceleration rate for the same time period in every cycle. When driving on a race track, the vehicle velocity changes continuously with constantly different acceleration and deceleration rates, depending on the race track.

Simulation can be used to optimize the brake disc and brake system design. The complex flow in and around the brake disc can be easily visualized and this makes it much easier to modify the design in order to optimize the brake cooling airflow, as will be shown in this paper. This will then lead to much lower brake system temperatures and makes it possible for the driver to push the limits even further during operation. In the same time, efficient brake cooling optimization can avoid large drag penalties, which can be a consequence of added brake cooling design changes like inefficient ducts and air deflectors. When the brake system is well cooled, the design could consider reducing the brake disc size in order to reduce the vehicle weight.

2 Set-up

2.1 Set-up Experiment

The experiment was performed on a vehicle race-track, where the front brake disc temperature was measured, together with vehicle speed. The test continued for many race track laps, until the brake disc temperatures reached a steady-state condition, meaning that the temperatures did not increase anymore after each lap. The ambient temperature was 26 °C.

2.2 Set-up Simulation

The simulation method is a coupled approach between the lattice Boltzmann based CFD solver PowerFLOW and the radiation/conduction solver PowerTHERM. This coupled approach has already been validated and applied in many publications [6–8]. The lattice Boltzmann based CFD solver is a particle based flow solver, where particle distribution functions move from grid point to grid point and each time collide with other particles. With mass and momentum conservation, the particle distribution functions obtain at each collision phase a new direction and velocity. It is therefore by nature an unsteady flow solver and the turbulence is solved by using a k-ε VLES turbulence model [9–22]. The wheel and brake disc are realistically rotated with a sliding mesh [23, 24].

The radiation/conduction solver is solved with the Crank-Nicholson implicit finite difference scheme [25, 26].

For the simulation set-up, all the vehicle geometry details are included in the CFD simulation. The vehicle geometry is shown in Fig. 1. There is a side pod that brings air through a channel to the brake disc. The brake geometry is shown in Figs. 2 and 3.

Fig. 1. Vehicle geometry used for the CFD simulation.

Fig. 2. Vehicle brake system geometry used for the CFD simulation.

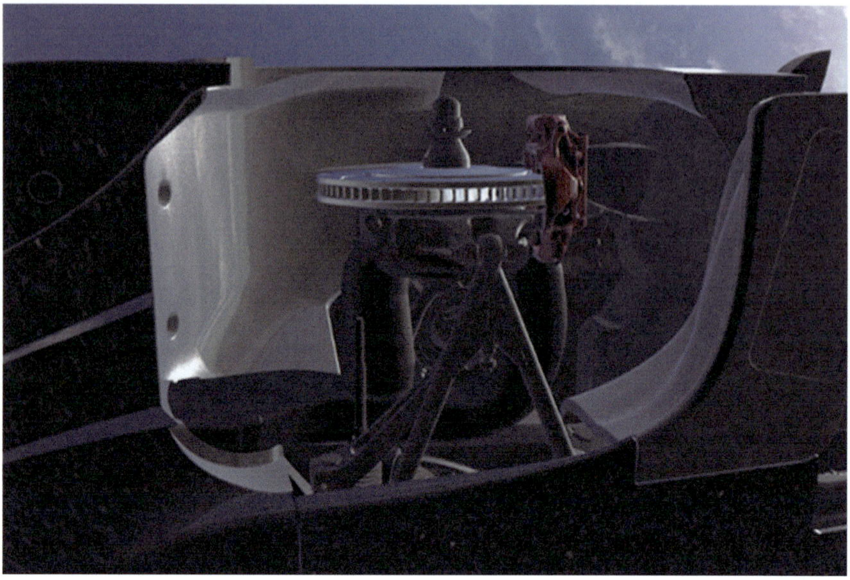

Fig. 3. Vehicle brake system geometry used for the CFD simulation, bottom view.

The duct splits up into two and part of the flow is guided to the front part of the brake disc and part of it to the rear, as shown in Fig. 4.

Fig. 4. Vehicle brake duct geometry used for the CFD simulation.

The complete vehicle geometry is simulated in the digital windtunnel at the maximum and minimum velocities from the race track. The wheels and brake discs are realistically rotated. The CFD volume mesh for the fluid domain is shown in Fig. 5.

Fig. 5. Fluid domain volume mesh used for the CFD simulation.

The CFD simulation is coupled to a radiation/conduction solver, which uses a 3D solid mesh for most brake system components, including the caliper internal geometry, like brake pads, pad plates, calipers pistons and the brake oil. Material and surface condition properties are taken from experiment. The coupled simulation process is shown in Fig. 6. The coupled CFD simulation provides the convection on an element basis. This is then used for calculating the race track duty cycle using the radiation/conduction solver on a stand-alone basis.

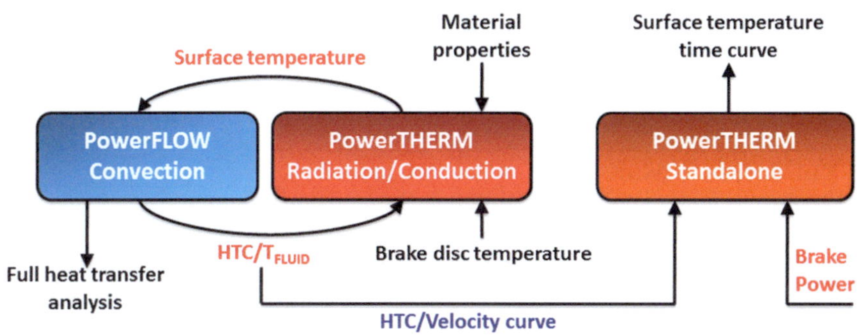

Fig. 6. Coupled simulation process.

The vehicle velocity as measured in experiment for one lap is shown in Fig. 7, together with the generated heat during braking. In the stand-alone radiation/conduction solver, the HTC's from the coupled CFD simulation are related to vehicle velocity and are used as a time dependent boundary condition. That means they vary over time per PID according to the vehicle velocity.

An additional input is the braking power, which is calculated from the vehicle kinetic energy transformation. This is also used as a heat input boundary condition for the stand-alone radiation/conduction simulation. That means that, at the moment of braking, the heat is inserted at the surface of the brake disc and brake pad as a function of time. The brake disc and pad conduction contact are also modeled as a function of time.

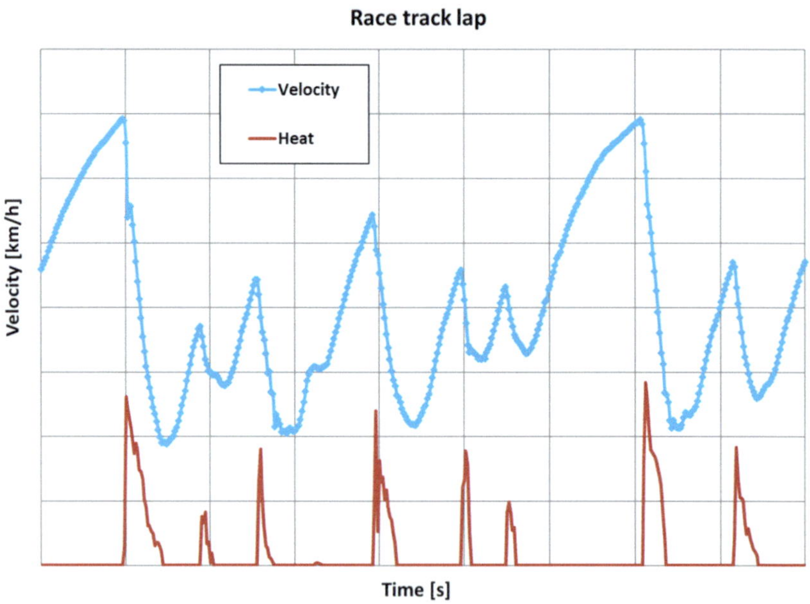

Fig. 7. Time dependent boundary conditions.

3 Validation Results

The brake disc duty cycle temperature during one lap is compared against experiment in Fig. 8. The results match quite nicely.

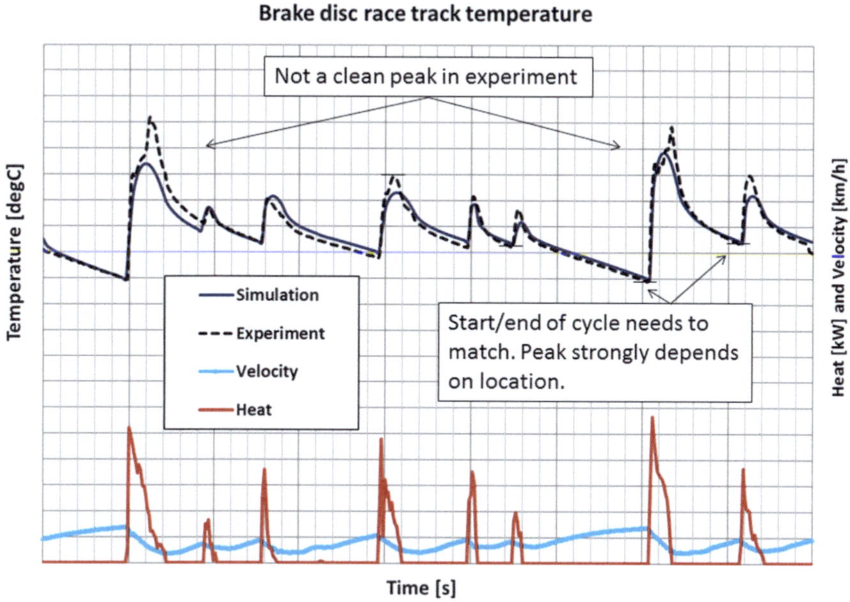

Fig. 8. Brake disc race track temperature.

As shown in the graph, the peak temperature is very difficult to compare. It is very sensitive to its location. During braking, the peak temperature rises rapidly within a few seconds. The highest peak temperatures are reached at the braking surface, because that is where the heat is generated. The left image in Fig. 9 shows the temperature during braking and it shows a very high temperature on the surface of the brake disc. This heat quickly spreads though the disc through conduction, causing strong temperature gradients through the disc. This is shown in the image on the right.

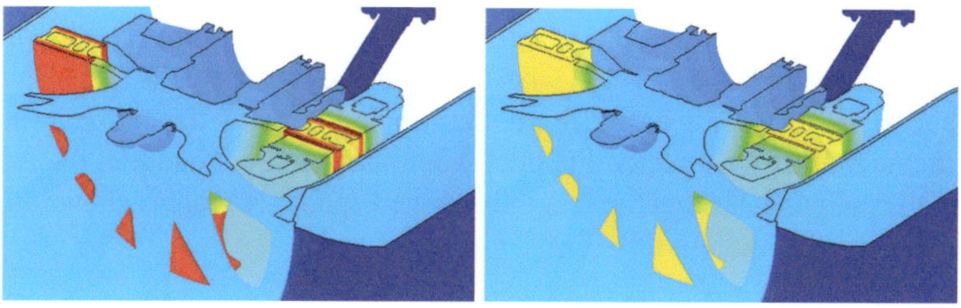

Fig. 9. Brake disc temperature during braking (left). At end of cooldown (right).

The peak temperature will strongly depend on location and therefore is difficult to compare. After the peak, during the cooldown phase of the disc, the heat spreads through the disc and at the end of the cooldown phase, the disc temperature becomes more uniform. That is when the temperature can be compared well against experiment. Considering one brake cycle, the starting temperature and end temperature should match. In between, a certain amount of heat is added due to braking and removed by convection, conduction and radiation, which leads to a new end temperature before the new braking cycle. If these two points match, then this is an indication that the amount of heat input, convection, conduction and radiation is calculated correctly.

Looking at the graph, most end points are calculated well and even the peak temperatures are not that far off. The first and second last peak are quite off, but the experimental data shows a bit of a strange shape, which might indicate a problem with the experiment. For such an aggressive experimental braking cycle where large amounts of energy are transformed into heat that is then absorbed by the brake discs in only a few seconds of time, it becomes difficult to obtain an accurate estimation of the exact heat-input that should be used in the simulation.

For example, the generated heat is absorbed by both the front and rear brakes. The amount that goes to the front is set by the OEM by adjusting the hydraulic pressure distribution settings. This is the so called brake distribution split. However, especially during such aggressive stops, this distribution might deviate from the default value, for a number of reasons, including weight distribution changes for the available tyre grip, the migration of the brake bias bar connecting the front and the rear pressure system and temperature effects. The thermodynamic phenomena here are very complex.

Taking into account all these effects, the simulation results compare reasonably well. What is relevant for the simulated results is that the averaged brake disc temperature level is predicted correctly. Minor deviations from the exact peak temperature should not be analyzed too critically.

4 Design Optimization

The developers of the vehicle wanted to investigate possibilities to lower the outboard oil temperature. The outboard side of the caliper is always more difficult to cool, because it is enclosed between the wheel and the disc. Most brake cooling ducts are located on the inboard side. They take air from the front of the vehicle or from underneath the vehicle and bring air to the inboard side of the brake system. For cooling the outboard side directly, the duct would have to go around all the brake disc or around the caliper and often there is only limited space for that, as shown in Fig. 10 for the considered vehicle in this study.

Fig. 10. Inboard side view of the wheel.

Figure 11 shows a duct design variant that brings more air to the caliper. Figure 12 shows that the variant brings more air all around the calipers which goes to the outboard side of the calipers. This will increase the convective heat transfer on the outboard side of the calipers.

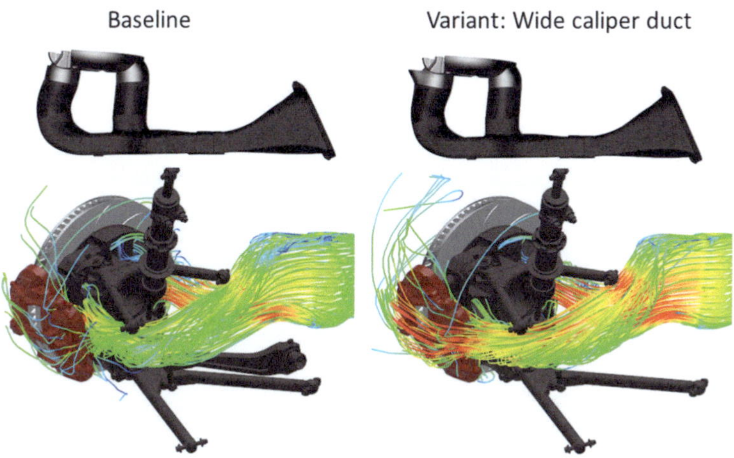

Fig. 11. Fluid velocity streamlines. Baseline (left) and variant (right). Inboard view.

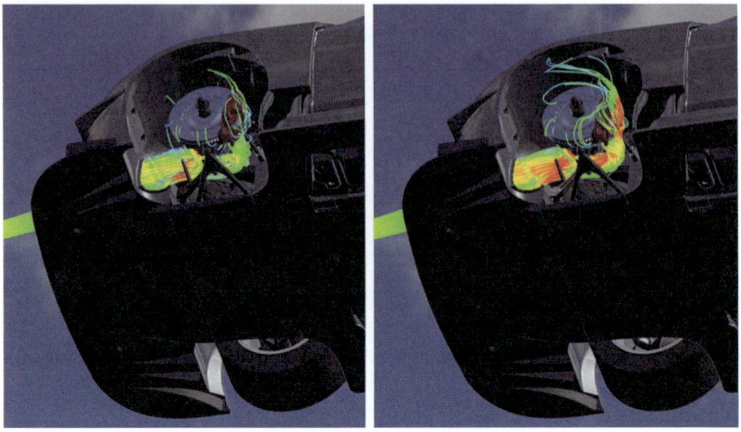

Fig. 12. Fluid velocity streamlines. Baseline (left) and variant (right). Outboard view.

Figure 13 shows the fluid velocity at a horizontal plane around the centre of the brake for the baseline and the variant. The front of the vehicle is on the right. The high velocity flow through the duct is clearly visible. The variant on the right shows that much more flow is going through the downstream duct. Also, more flow is passing around the caliper, both indicated with a black arrow.

The increase in velocity in the downstream duct is caused by the reduction in stagnation pressure after opening up the downstream duct, which is shown in Fig. 14. The flow stagnates less and this increases the massflow.

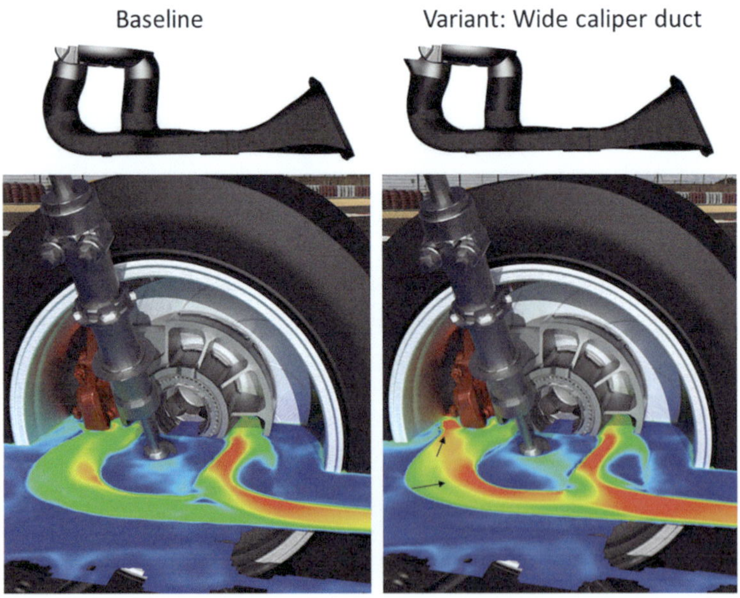

Fig. 13. Horizontal fluid velocity plane at Z = 0.3 m.

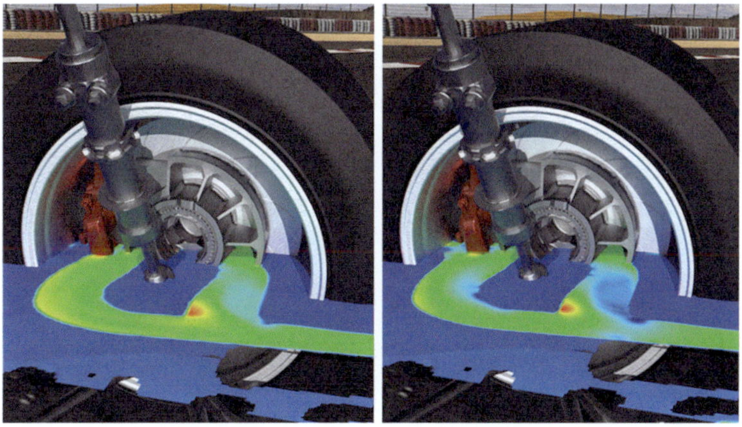

Fig. 14. Horizontal fluid static pressure plane at Z = 0.3 m.

The surface HTC's are shown in Fig. 15 (outboard) and Fig. 16 (inboard). The impact of the flow on the outboard side of the calipers can also be seen from the surface HTC values. The variant shows increased surface HTC on the outboard side of the calipers, indicated with a black arrow, see Fig. 15. The inboard HTC's are shown in Fig. 16 and we see that they are not much impacted by the design change. Table 1 shows that the brake disc HTC's are only slightly lower, but that the outboard caliper side has much higher HTC's.

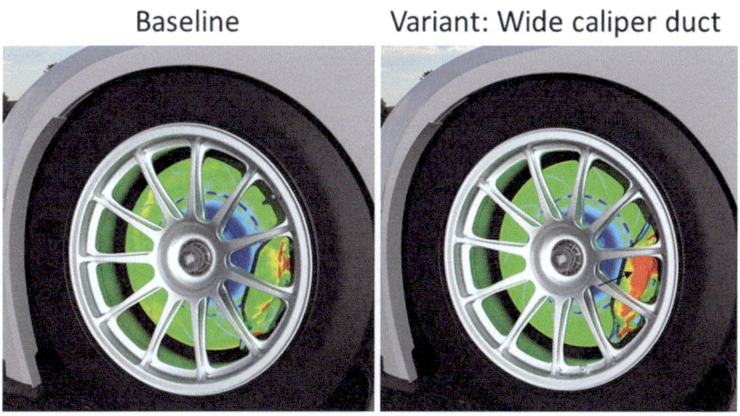

Fig. 15. Surface HTC distribution, outboard

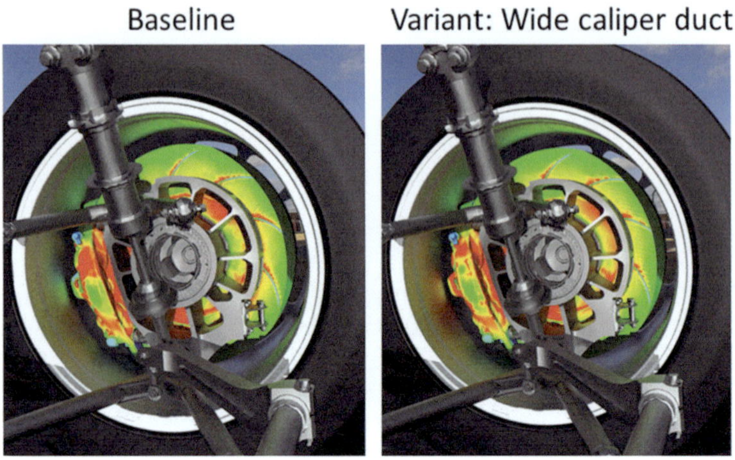

Fig. 16. Surface HTC distribution, inboard

Table 1. Surface HTC per part. Percentage of difference compared to the baseline.

HTC [W/m2K]	Braking surface IB	Braking surface OB	Brake vents	Brake hat IB	Caliper OB
Variant compared to Baseline	-4.5%	-7.1%	-4.8%	-5.2%	+26.0%

The brake disc temperature and the outboard brake oil temperature for the variant during the race track lap are shown in Fig. 17. We can see that the brake disc temperature is slightly higher, however, the outboard brake oil temperature is clearly lower, which was the main objective. Since the brake oil has a much lower critical temperature than the brake disc, the temperature reduction on the brake oil has a much more significant impact than the temperature increase for the brake disc.

Finally, the simulation makes it possible to easily look at the massflow through different parts of the brake duct, shown in Table 2. In total, there is more massflow through the duct. This can be explained by the fact that when opening up the downstream duct, the pressure and blockage of the flow becomes less and more flow can enter. It also causes more flow to exit through the downstream duct, which has the consequence that less flow goes through the upstream duct. However, as indicated previously, this reduces only slightly the brake disc cooling.

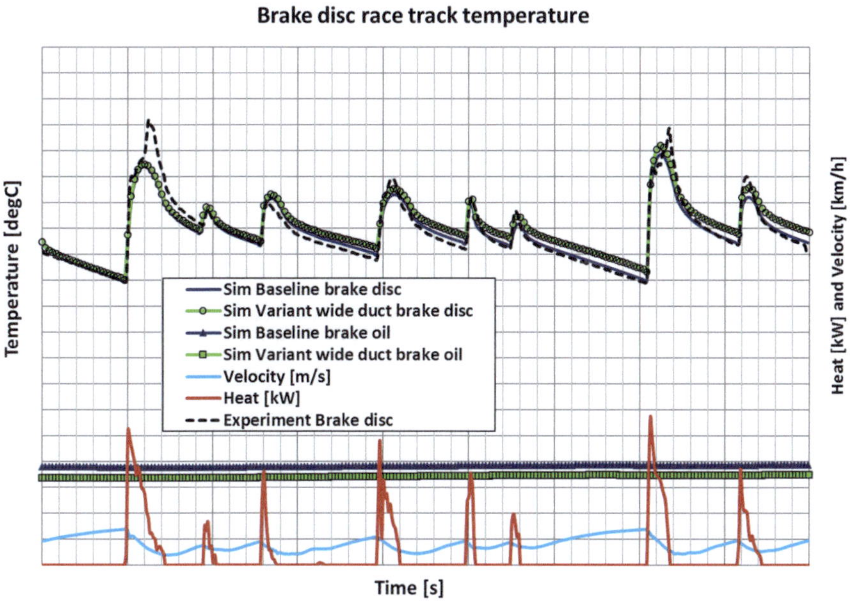

Fig. 17. Brake system temperatures during race track lap. Baseline and variant.

Table 2. Brake duct massflow. Percentage of difference compared to the baseline.

massflow [kg/s]	Entry duct	Upstream duct	Downstream duct	Brake disc
Variant compared to Baseline	+12.9%	-7.7%	+35.3%	-7.5%

5 Future Work

In this study the calipers outboard oil temperature was reduced, but the compromise was a higher brake disc temperature. As a follow-up study, the shape of brake duct itself can be further optimized in order to increase the massflow, for example by trying to reduce the flow separations inside the duct. With higher massflow through the duct, both the calipers and the brake disc temperature will be reduced instead of compromising one or the other.

6 Conclusions

The coupled simulation methodology is able to predict the brake disc temperatures during a race track lap well for the considered validation vehicle. The study shows a design variant that is able with a small design modification to lower the caliper outboard brake oil temperature, without sacrificing the brake disc cooling and temperature too much. Since the design change was rather small, drastic alterations were avoided like a much bigger brake cooling duct opening or a much bigger brake disc in order to lower the brake system temperatures. Such changes lead to significant drag and/or vehicle weight penalties. Simulation made it possible to optimize the design virtually. The race track temperatures are calculated with the radiation/conduction solver on a stand-alone base, using the convection from the CFD solver as a boundary condition. Realistic rotation of the brake disc and wheel is used and a 3D solid mesh for conduction inside the brake system. The study shows that the coupled simulation method can be used as a predictive tool to optimize the vehicle. Simulation provides the additional benefit that two aerodynamic designs can be compared directly to each other for the same drive cycle. The results will not depend on the driver, the driver brake reaction reproducibility or weather conditions. The flow can be visualized, which makes it possible to efficiently improve the brake disc cooling airflow, without significant aerodynamic drag penalties. The analysis can be done in early design stages, before real prototypes are built and can avoid expensive design changes in later design phases.

References

1. Schütz, T., Wiedemann, J., Wickern, G., Mukutmoni, D., Wang, Z., Alajbegovic, A.: A coupled approach to brake cooling simulation. In: Mira Conference (2008)
2. Mukutmoni, D., Jelić, S., Han, J., Haffey, M.: Role of accurate numerical simulation of brake cooldown in brake design process. In: SAE 2012 Brake Colloquium & Exhibition, SAE 2012-01-1811, San Diego, US (2012)
3. Jelić, S., Meyland, S., Jansen, W., Alajbegovic, A.: A coupled approach to brake duty cycle simulation. In: 8th MIRA International Vehicle Aerodynamics Conference, Oxfordshire, UK (2010)
4. Meachair, D., Elliot, G., Jelić, S., Fechner, B., Bhambare, K.: Brake duty cycle simulation for thermal design of vehicle braking system. In: Eurobrake 2013, EB2013-MS-003, UK (2013)
5. Sun, S., Liao, G., Fu, Q., Lu, K., Zhao, J., Li, Z., Chen, J., Shi, G., Jelić, S., Li, B.: A coupled approach to truck drum brake cooling. SAE Technical Paper 2015-01-2901, Chicago, US (2015).https://doi.org/10.4271/2015-01-2901
6. Jelić, S., Fares, E., Alajbegovic, A.: Lattice-Boltzmann flow simulation of the modified sae model with heated plug including conduction and radiation effects. In: Vehicle Thermal Management Systems 8, Nottingham, UK, 2007
7. Mukutmoni, D., Han, J., Alajbegovic, A., Colibert, L., Helene, M.: Numerical simulation of transient thermal convection of heated plate. In: SAE World Congress, SAE 2010-01-0550, Detroit (2010)

8. Mukutmoni, D., Alajbegovic, A., Han, J.: Numerical simulation of transient thermal convection of a full vehicle. In: SAE World Congress, SAE 2011-01-0645, Detroit, USA (2011)
9. Frisch, U., Hasslacher, B., Pomeau, Y.: Lattice-gas automata for the navier-stokes equations. Phys. Rev. Lett. **56**, 1505–1508 (1986)
10. Chen, H., Chen, S., Matthaeus, W.: Recovery of the Navier-Stokes equations using a lattice-gas Boltzmann method. Phys. Rev. A **45**, 5339–5342 (1992)
11. Chen, H., Teixeira, C., Molvig, K.: Digital physics approach to computational fluid dynamics: some basic theoretical features. Int. J. Mod. Phys. C **9**(8), 675 (1997)
12. Chen, H., Teixeira, C., Molvig, K.: Realization of fluid boundary conditions via discrete Boltzmann dynamics. Int. J. Mod. Phys. C **9**(8), 1281–1292 (1998)
13. Thantanapally, C., Singh, S., Succi, S., Ansumali, S.: Quasi-equilibrium lattice Boltzmann models with tunable Prandtl number for incompressible hydrodynamics. Int. J. Mod. Phys. C **24**(12), 1340004 (2013)
14. Chen, H.: Volumetric formulation of the lattice boltzmann method for fluid dynamics: basic concept. Phys. Rev. E **58**, 3955–3963 (1998)
15. Shan, X., Chen, H.: A general multiple-relaxation-time Boltzmann collision model. Int. J. Mod. Phys. C **18**, 635 (2007)
16. Chen, H., Goldhirsh, I., Orszag, S.: Discrete rotational symmetry, moment isotropy, and higher order lattice Boltzmann models. J. Sci. Comput. **34**, 87–112 (2007)
17. Bhatnagar, P., Gross, E., Krook, M.: A model for collision processes in gases. I. small amplitude processes in charged and neutral one-component system. Pys. Rev. **94**, 511–525 (1954)
18. Chapman, S., Cowling, T.: The Mathematical Theory of Non-uniform Gases. Cambridge University Press, Cambridge (1990)
19. Chen, H., Kandasamy, S., Orszag, S., Shock, R., Succi, S., Yakhot, V.: Extended Boltzmann kinetic equation for turbulent flows. Science **301**, 633–636 (2003)
20. Shan, X., Chen, H.: Lattice Boltzmann model for simulating flows with multiple phases and components. Phys. Rev. E **47**, 1815–1819 (1993)
21. Li, Y., Shock, R., Zhang, R., Chen, H.: Numerical study of flow past an impulsively started cylinder by lattice boltzmann method. J. Fluid Mech. **519**, 273–300 (2004)
22. Chen, H., Orszag, S., Staroselsky, I., Succi, S.: Expanded analogy between Boltzmann kinetic theory of fluid and turbulence. J. Fluid Mech. **519**, 307–314 (2004)
23. Zhang, R., et al.: Lattice Boltzmann approach for local reference frames. DSFD-17 Special Edition, Communications in Computational Physics (2011)
24. Guo, Z., Zhen, C., Shi, B.: Discrete lattice effects on the forcing term in the lattice Boltzmann method. Phys. Rev. E **65**, 046308 (2002)
25. PowerTHERM technical documentation. Version 7.1. Thermoanalytics, Calumet, US
26. Strauss, W.: Partial Differential Equations, An Introduction. Wiley, Hoboken (2008)

A New Process to Evaluate the Risk of an Engine Power Drop Caused by Snow Particles

Christoph Huber[1(✉)], Heinrich Reister[1], Thomas Binner[1], and Bernhard Weigand[2]

[1] Daimler AG, Benzstrasse, Tor 16, 059-X687, 71059 Sindelfingen, Germany
christoph.c.huber@daimler.com
[2] ITLR, University of Stuttgart, Stuttgart, Germany

Abstract. In cold and snowy regions roads are often covered with snow during the winter time. On these roads snow particles can be rolled up by vehicles and these particles can penetrate into the air intake system of the engine. As a result an engine power drop can occur. This paper discusses three different possibilities to evaluate the risk of such an engine power drop. One possibility is given by measurements on a natural testing area. The second possibility consists of measurements in a climatic wind tunnel and the third possibility is given by numerical simulations. As examples for the experimental testing a natural testing area in Arjeplog (Sweden) and two climatic wind tunnels in and near Stuttgart will be explained in more detail. Information for the snow particle generation, the snow particle size and the snow mass flux will be provided. Additionally to this experimental testing a method for the virtual testing will be presented. This method is based on an Eulerian/Lagrangian approach within a commercial CFD-software. The simulation method allows an approximate calculation of the snow amount, which penetrates into the air filter body.

In the past prototypes were built and measurements were carried out to evaluate the snow amount, which penetrates into the air filter body. These measurements were possible at a later stage of the vehicle development process. We suggest to introduce the simulation method in an earlier stage of the development process. Thereby the air intake system can reach a high level of maturity before first prototypes are built.

1 Introduction

In cold and snowy regions (e.g., Northern Europe) roads are often covered with snow during the wintertime. On such roads snow particles can be rolled up by vehicles (as shown in Fig. 1). When vehicles are driven in a row, snow particles can be rolled up by vehicles in front and penetrate into the air filter body of vehicles behind. Within the air filter body of the engine the snow particles are separated from the air flow by the air filter. Depending on the amount of snow within the air filter body an engine power drop can occur. Even an engine stop is possible.

Traditionally there are two possibilities to evaluate the risk of an engine power drop, which is caused by snow particles:

J. Wiedemann (ed.), *Progress in Vehicle Aerodynamics and Thermal Management*,
https://doi.org/10.1007/978-3-319-67822-1_2

Fig. 1. Snow particles rolled up by a vehicle.

1. Automobile manufacturers carry out measurements on natural testing areas. On such testing areas vehicles can be driven in a row (one behind the other). After a certain time, the snow amount inside the air filter body and the engine power drop can be evaluated.
2. Automobile manufacturers carry out measurements in climatic wind tunnels. Snow is artificially produced to create environmental conditions, which are similar to those on roads and natural testing areas. As on the natural testing area the snow amount inside the air filter body and the engine power drop can be evaluated after the measurement.

However, both possibilities have one large disadvantage: Prototypes of the vehicle have to be built to carry out such measurements. To develop new vehicles in a cost-effective manner, automobile manufacturers use numerical simulations to achieve a high level of maturity of the vehicle before first prototypes are built. Exemplarily a reduced sketch of the vehicle development process of the Daimler AG is shown in Fig. 2. This development process starts with the virtual testing of the vehicle. During this virtual testing the vehicle is improved until a certain level of maturity is achieved. Having achieved this level, prototypes are built and the testing is continued in wind tunnels, on natural testing areas and on roads. In the past the virtual testing didn't provide the possibility to evaluate the risk of an engine power drop, which is caused by snow particles. To evaluate this risk, measurements were carried out at a later stage of the vehicle development process. If adjustments to the vehicle were required in this late stage, high costs could emerge. To evaluate the risk of an engine power drop during the virtual testing phase, we suggest to use the simulation method as described in [1]. Using this simulation method the amount of snow, which penetrates into the air filter body, can be predicted and the air intake system can be improved at an early stage of the vehicle development process.

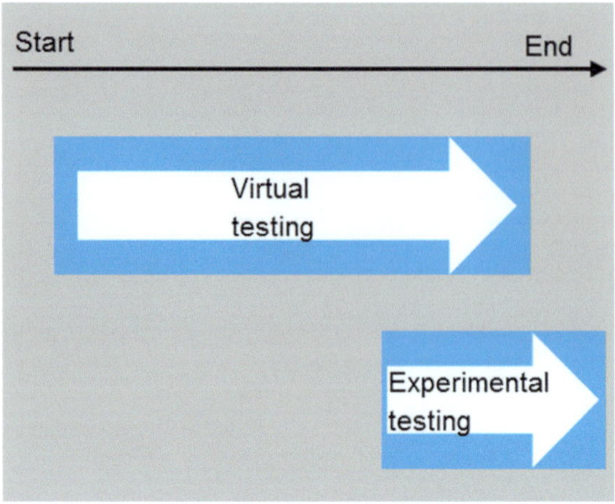

Fig. 2. Reduced sketch of the vehicle development process of the Daimler AG.

In the next section the traditional possibilities to evaluate the risk of an engine power drop are shown. After that the simulation method, which allows a virtual assessment of the risk, is explained. At the end results of the virtual testing will be compared to results of the experimental testing.

2 Traditional Possibilities to Evaluate the Risk of a Power Drop

To evaluate the risk of an engine power drop, which is caused by snow particles, Daimler AG has different testing areas in Northern Europe and two climatic wind tunnels in and near Stuttgart. In the following sections a testing Area in Arjeplog (Sweden) and the two climatic wind tunnels will be introduced.

2.1 Testing Area in Arjeplog

During the winter time Arjeplog offers cold temperatures, frozen lakes and a lot of snow. Due to these conditions Arjeplog is well suited for winter tests. The testing area of Daimler AG is on a frozen lake, which provides a circuit for the testing. To carry out the tests, two vehicles are required. Vehicle one drives in front to roll up snow (see Fig. 1). Vehicle two (the test vehicle) follows vehicle one in a certain distance. At the end of the measurement the snow amount inside the air filter body and a possible engine power drop can be evaluated.

In nature snow crystals can have many different particle shapes. For example there are needles, dendrites, plates or columns [2, 3]. These snow crystals grow from water vapor [3]. The size of these snow crystals in clouds are typically between a few micrometers and one centimeter [4]. Some of these snow crystals are hollow. Heymsfield [5] and Ono [6] carried out measurements to determine the snow crystal density.

In their measurements values between approximately 500 kg/m^3 and 917 kg/m^3 were observed. In this work rolled up snow particles are of interest. It is likely that these rolled up snow particles consist of fragments of snow crystals and snowflakes. In [1] the particle size of these rolled up snow particles was investigated. As can be seen in Fig. 3 the rolled up snow particles are typically between 30 μm and 500 μm. Figure 3 includes a distribution for the particle counts and for the volume of the particles. To measure the snow particle size a so called snow particle counter from Niigata Electric was used (see Fig. 4). The particle size distribution was measured in a distance of 0.7 m to the ground. In [1] the volume median diameter was used to characterize the particle size distribution. For the distribution shown in Fig. 3 the volume median diameter is approximately 260 μm. In [1] this volume median diameter was measured on different days on the testing area. Values between 180 μm and 380 μm were observed. This scattering represents a challenge for the natural testing area with respect to reproducibility: Fresh snow, different temperatures or different winds influence the measurement.

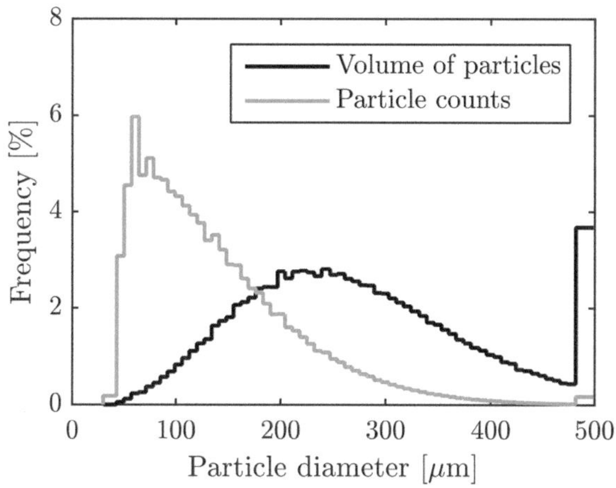

Fig. 3. Typical particle size distribution of rolled up snow on the test area in Arjeplog; data from [1].

Additionally to the snow particle size the snow mass flux in front of the test vehicle was measured in [1]. As for the particle size the snow particle counter was used to measure the mass flux. The results are shown in Fig. 5. On different days values between 20 and 240 g/(m^{2}s) were observed. As mentioned before this scattering occurs due to different ambient conditions. Since the snow mass flux in front of the test vehicle has a direct influence on the snow amount inside the air filter body of the test vehicle, the normalized snow mass flux n was introduced in [1]:

$$n = \frac{snow\ mass\ flux\ at\ the\ inlet\ of\ the\ air\ filter\ body}{snow\ mass\ flux\ in\ front\ of\ the\ vehicle}. \tag{1}$$

Fig. 4. Test vehicle with snow particle counter.

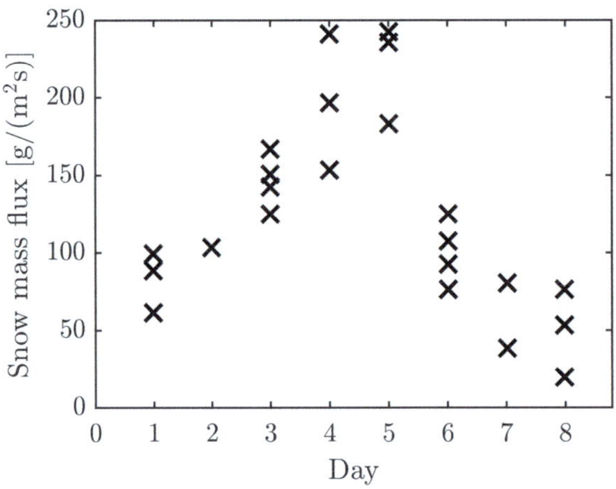

Fig. 5. Snow mass flux in front of the test vehicle on the testing area in Arjeplog at different days; data from [1].

This ratio compares the snow mass flux at the inlet of the air filter body to the snow mass flux in front of the test vehicle. The snow mass flux at the inlet of the air filter body was determined by weighing the snow amount inside the air filter body and dividing this amount by the measuring duration and the inlet area of the air filter body. The snow mass flux in front of the test vehicle mas measured with the help of the snow particle counter. The idea of introducing the ratio n is to increase the comparability between different measurements.

2.2 Climatic Wind Tunnels

The large advantage of a climatic wind tunnel is that measurements can be carried out all year long and that the conditions are reproducible. However, there is one difference (regarding the snow particles) compared to the testing on a natural testing area: the snow particles are artificially produced. Often small water droplets are generated in climatic wind tunnels. These water droplets will freeze in the wind tunnel and become solid. So these snow particles consist of frozen water droplets. Unlike to natural snow crystals these artificially produced snow particles don't grow from water vapor and this is the reason why the snow particle shape is different.

2.2.1 Climatic Wind Tunnel in Sindelfingen

One climatic wind tunnel of Daimler AG, which can generate snow particles, is located in Sindelfingen. In this wind tunnel temperatures between −40 °C and 40 °C can be realized. The maximum wind speed is 265 km/h. Figure 6 shows a schematic drawing of this wind tunnel. For snow tests two snow wings can be inserted into the nozzle of the climatic wind tunnel (see Fig. 6). Each wing has 8 two-fluid nozzles to atomize water and generate small droplets. As can be seen in the figure, the water is injected against the flow direction. Due to this procedure the particle flight time, until the particle impinges on the vehicle, can be increased. Especially for larger water droplets this increased flight time is necessary to obtain completely frozen particles.

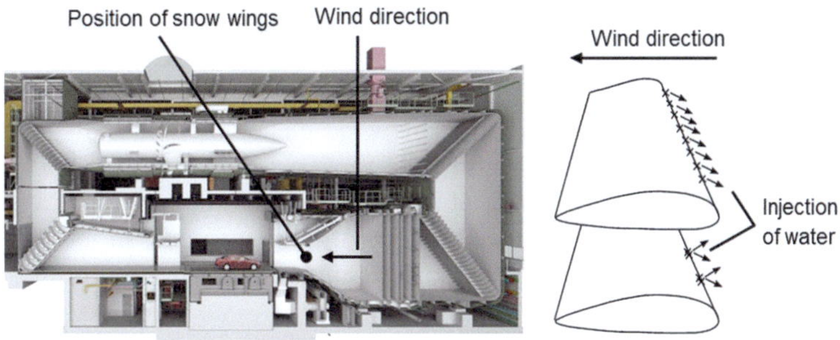

Fig. 6. Schematic drawing of the climatic wind tunnel in Sindelfingen; two snow wings can be inserted to generate snow particles.

In Fig. 7 the size distribution of these artificially produced snow particles is shown. As can be seen the particles are between 10 μm and 500 μm. The volume median diameter for this distribution is 85 μm. For this measurement the device Spraytec from Malvern Instruments Ltd. was used. Since the water droplets require a certain time to completely freeze the snow particle size in the climatic wind tunnel is limited. A typical value for the snow mass flux in this wind tunnel is approximately 70 g/(m^{2}s).

To evaluate the risk of an engine power drop in the climatic wind tunnel the two snow wings are inserted into the wind tunnel and after a certain time the snow amount inside the air filter body of the test vehicle and the engine power drop are evaluated.

2.2.2 Climatic Wind Tunnel in Stuttgart

Another climatic wind tunnel of Daimler AG, which can generate snow particles, is in Stuttgart. This wind tunnel is shown in Fig. 8. The maximum wind speed for this wind tunnel is 200 km/h. To generate snow particles a snow canon can be inserted into the wind tunnel. This snow canon produces small water droplets, which will freeze due to

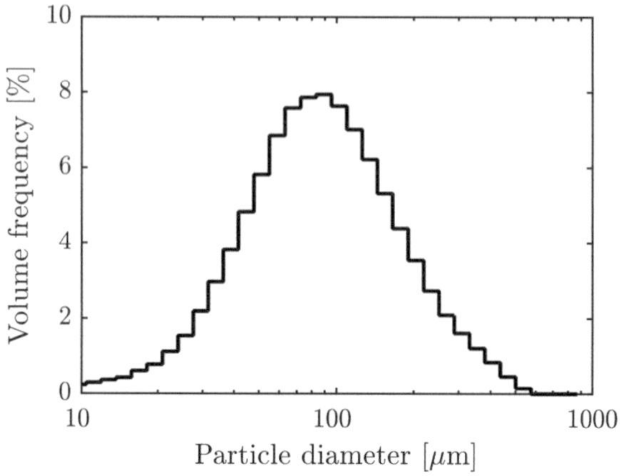

Fig. 7. Snow size distribution in the climatic wind tunnel in Sindelfingen; data from [7].

the low temperature in the climatic wind tunnel. As indicated in Fig. 8 the particles are sprayed into an opening. From this point the particles follow the air flow and eventually impinge on the vehicle. Using the snow particle counter a volume median diameter of approximately 160 μm was observed for this wind tunnel [1]. The snow particle size distribution is shown in Fig. 9. Regarding the snow mass flux a value of 105 g/(m^{2}s) was observed [1]. To evaluate the risk of an engine power drop in this wind tunnel the snow canon is activated and after a certain time the snow amount inside the air filter body of the test vehicle and the engine power drop are evaluated.

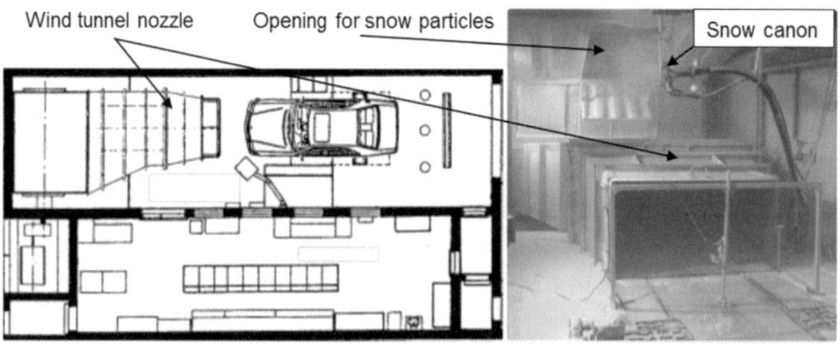

Fig. 8. Snow production in the climatic wind tunnel in Stuttgart.

3 Simulation Approach to Calculate the Snow Ingress

In [1, 7] a simulation approach to predict the amount of snow, which penetrates into the air filter body of the engine, is introduced. This simulation method allows a virtual testing of the vehicle during an early stage of the development process.

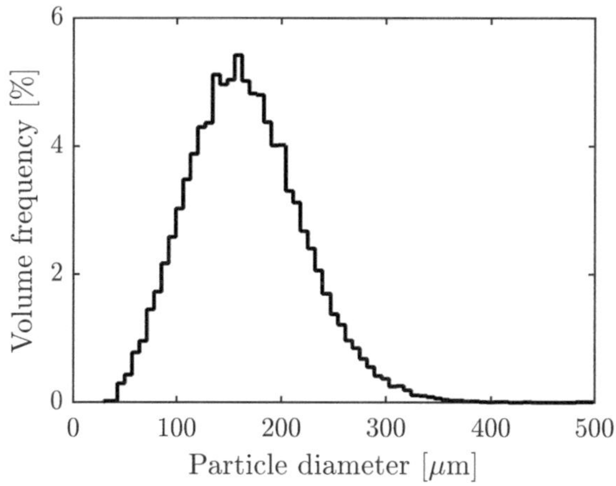

Fig. 9. Snow size distribution in the climatic wind tunnel in Stuttgart.

The simulation method for the particle-laden flow is based on an Eulerian/Lagrangian framework within a commercial CFD-software. For the continuous phase (air) an Eulerian framework is applied and for the dispersed phase (snow particles) a Lagrangian framework is utilized. By solving the Reynolds-averaged Navier-Stokes (RANS) equations the air flow can be calculated. For the particle-laden flow a one-way coupled flow is assumed. So the dispersed phase doesn't affect the continuous phase and particle-particle interactions are neglected. Furthermore the parcel concept is used. A parcel represents a group of particles, which have similar properties [8]. To calculate the snow particle trajectories Newton's second law has to be solved for each representative particle (or parcel):

$$m_p \frac{d\boldsymbol{v}_p}{dt} = \boldsymbol{F}. \tag{2}$$

Using this law the change of the particle velocity $\boldsymbol{v}_p$ can be determined by calculating the forces $\boldsymbol{F}$, which are acting on the particle. In Eq. (2) m_p is the mass of the particle. In the simulation approach described in [1] the drag force

$$\boldsymbol{F}_d = \frac{1}{2}\rho A_p c_d \left(\boldsymbol{v} - \boldsymbol{v}_p\right)\left|\boldsymbol{v} - \boldsymbol{v}_p\right|, \tag{3}$$

the gravitational force

$$\boldsymbol{F}_g = m_p \boldsymbol{g}, \tag{4}$$

and the pressure gradient force [8]

$$\boldsymbol{F}_p = -V_p \nabla p_{static} \tag{5}$$

are taken into account.

In Eq. (3) ρ denotes the fluid density, A_p the cross-sectional area of the particle, c_d the drag coefficient and $\boldsymbol{v}$ the velocity of the fluid. In Eq. (4) $\boldsymbol{g}$ is the gravitational acceleration and in Eq. (5) V_p denotes the volume of a particle and ∇p_{static} is the gradient of the static pressure in the continuous phase. Using these forces the equation of motion results in

$$m_p \frac{d\boldsymbol{v}_p}{dt} = \boldsymbol{F}_d + \boldsymbol{F}_g + \boldsymbol{F}_p. \tag{6}$$

3.1 Drag Force

To determine the drag force $\boldsymbol{F}_d$, the snow particles are treated as isolated particles in [1]. So the drag force of a particle is not influenced by its neighbors. To simulate spherical particles (for example frozen droplets or fragments of snow crystals, which are similar to a sphere) a correlation from Schiller and Naumann [9] is used to determine the drag coefficient c_d. To simulate non-spherical snow particles (e.g., plates, columns, needles) correlations from Loth [10] are used for c_d. Further information about the drag coefficient is given in [1].

3.2 Wall-Impact Behavior

Before snow particles enter the air filter body, they collide with vehicle parts such as the radiator grille or the license plate. As explained by Sommerfeld [11] particle-wall collisions can considerably influence the particle motion. Especially for larger particles this influence is significant. The collision process depends on many parameters. Some of them are: Particle incidence angle, particle velocities before collision (translational and rotational), properties of the particle and the wall materials, particle shape and the surface roughness of the wall [11].

When ice particles collide with a cold wall (<0 °C), there are two possible outcomes [12, 13]:

- they bounce off without fragmentation or
- they bounce off and disintegrate.

Hauk et al. [14] introduced a map for the ice particle breakup. Depending on the particle diameter and the particle velocity before the collision this map allows to estimate if fragmentation occurs. Therewith, we conclude that for typical driving speeds of 50 to 80 km/h on snowy roads and rolled up snow particles, with a size ranging from 10 μm to 500 μm, major fragmentation doesn't occur. For this reason fragmentation is not considered in the simulation approach described in [1].

To calculate the particle velocity after the wall collision an approach based on the momentum equations for the collision of a spherical particle with a wall [11, 15, 16] is used. Furthermore, it is assumed that the particles rebound on a smooth surface and that the particle rotation can be neglected. Using this approach the following equations are obtained:

$$\text{if } \left|\frac{v_{p,n}^{(1)}}{v_{p,t}^{(1)}}\right| \geq \frac{2}{7(1+e_n)\mu_k}$$

$$v_{p,n}^{(2)} = -e_n v_{p,n}^{(1)} \tag{7}$$

$$v_{p,t}^{(2)} = \frac{5}{7} v_{p,t}^{(1)} \tag{8}$$

$$\text{and if } \left|\frac{v_{p,n}^{(1)}}{v_{p,t}^{(1)}}\right| < \frac{2}{7(1+e_n)\mu_k}$$

$$v_{p,n}^{(2)} = -e_n v_{p,n}^{(1)} \tag{9}$$

$$v_{p,t}^{(2)} = (1 - \mu_k(1+e_n)\tan(\theta))v_{p,t}^{(1)}. \tag{10}$$

In these equations $v_{p,n}$ denotes the component of the particle velocity normal to the wall and $v_{p,t}$ is the component of the particle velocity tangential to the wall. The superscript (1) refers to the velocity before the collision and the superscript (2) to the velocity after the collision. θ is the particle incidence angle, which is defined as the angle between the particle velocity vector and the tangential vector of the wall. For the coefficient of kinetic friction μ_k a value of 0.05 is used [17]. To estimate the restitution coefficient e_n a model from Villedieu et al. is used [18]. Further information about the simulation approach can be found in [1, 7].

4 Comparison Between Simulation Results and Measurements

In this section the snow mass flux at the inlet of the air filter body, which was obtained by using the presented simulation approach, will be compared to measurements, which were carried out on the testing area in Arjeplog and in the climatic wind tunnel in Stuttgart. These results were first shown in [1].

4.1 Simulation Set-up

The simulations were realized in STAR-CCM+ [8]. As shown in Fig. 10 a full vehicle model was considered. The snow particles were injected in a distance of about 1 m in front of the vehicle. Regarding the snow particle size, the distribution from Fig. 3 was used. For the simulations and the measurements the normalized snow mass flux n, which compares the snow mass flux at the inlet of the air filter body to the snow mass flux in front of the test vehicle, was evaluated (see Eq. (1)).

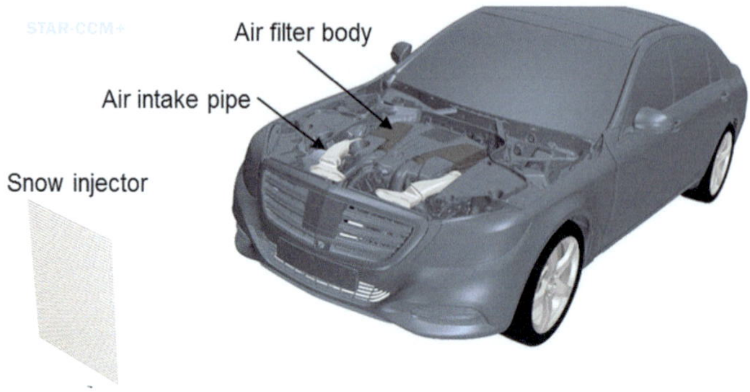

Fig. 10. Test vehicle with snow injector.

4.2 Results

Figure 11 compares the normalized snow mass flux n between two different engine speeds: 1000 rpm and 4500 rpm. Results are shown for the measurements carried out on the testing area in Arjeplog and for the simulations. In the simulations different values for the snow particle density were investigated since snow crystals can be hollow. In nature values between 500 kg/m^3 and 917 kg/m^3 can be observed [5, 6]. For this reason the values 917 kg/m^3, 750 kg/m^3 and 500 kg/m^3 were investigated. For the investigation of the snow particle density a spherical shape was assumed. As mentioned before natural snow crystals can have different particle shapes. Since the particle shape is not known for the testing area in Arjeplog, different basic shapes were considered: a sphere, a thin plate, a thick plate, a column and a needle. For the investigation of the particle shape a density of 917 kg/m^3 was assumed. As shown in Fig. 11, a lower engine speed results in a lower snow amount inside the air filter body. The comparison

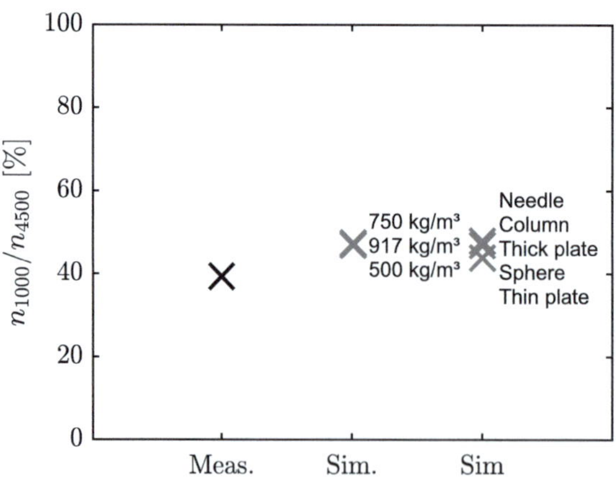

Fig. 11. Normalized snow mass flux n for the measurement (Meas.) and the simulations (Sim.): illustrated is the ratio of an engine speed of 1000 rpm and 4500 rpm; data from [1].

between the measurements and the simulation results shows a good agreement. It is likely that the difference between the simulation results and the measurements occur due to the modeling of the wall-impact behavior of snow particles. In the simulations it is assumed that all particles rebound on a smooth surface. However, the air intake pipe of the test vehicle is made of a felt-like material. During the tests in Arjeplog it was observed, that a small amount of snow can stick to this felt-like material. Furthermore, the restitution coefficient e_n can only be estimated by the model in [18].

Figure 12 shows the normalized snow mass flux n for two different vehicle configurations. As shown in this figure, configuration two has a radiator grille, which is partially closed. Illustrated are the simulation results and the measurements from Arjeplog. As mentioned before different snow particle shapes and different values for the snow particle density were investigated in the simulations. Especially the particle shape can be important for the simulation. Unfortunately this parameter is not known for the testing area in Arjeplog. Furthermore, it is possible that the particle shape varies on the natural testing area and on roads at different days. So during the virtual testing different particle shapes should be considered.

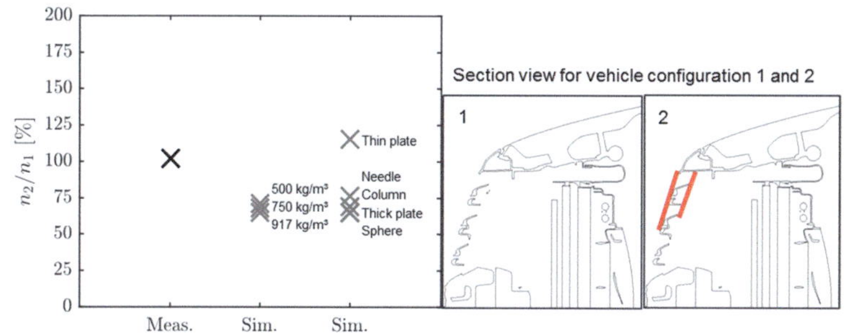

Fig. 12. Normalized snow mass flux n for the measurement (Meas.) and the simulations (Sim.): Comparison for two different vehicle configurations; data from [1].

In Fig. 13 the normalized snow mass flux n of a vehicle with an improved air intake system is compared to a reference vehicle. As shown in the figure the vehicle with the improved air intake system has a different air intake pipe. This new air intake pipe was designed to be advantageous for all considered snow particle shapes. In Fig. 13 the simulation results are compared to measurements, which were carried out in the climatic wind tunnel in Stuttgart. As shown in the figure the snow amount inside the air filter body could be reduced by 95% in the climatic wind tunnel. The simulation model predicted a similar benefit (99%). More information about the improved air intake system can be found in [1]. It should be noted that the new air intake system was designed with regard to snow. When designing the air intake system of a new vehicle there are two additional objectives: the total pressure drop and the risk of a hydro-lock should be low.

As these comparisons show, the simulation method can be used for an approximate calculation of the snow ingress. Furthermore, the simulation method can be used for

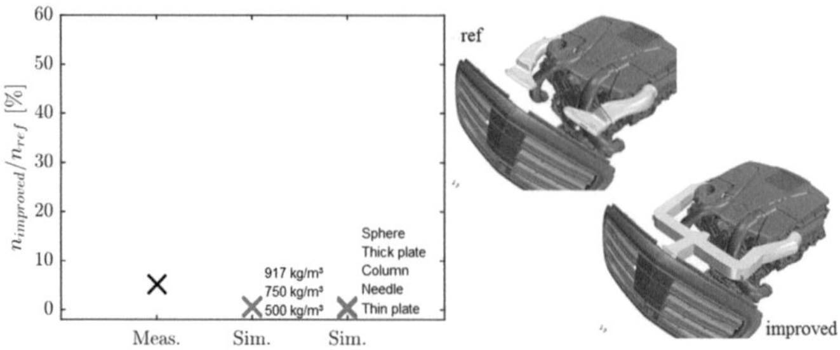

Fig. 13. Normalized snow mass flux n for the measurement (Meas.) and the simulations (Sim.): Comparison between an improved system and the reference system; data from [1].

improving the air intake system. To develop new vehicles in a cost-effective manner we suggest to introduce the simulation method within the virtual testing stage. During this stage the snow amount inside the air filter body of a successor model can be compared to its predecessor. If necessary the air intake system can be improved at an early stage of the vehicle development process and a high level of maturity can be achieved before first prototypes are built. Since there are simplifications applied in the simulation model (e.g., the wall-impact behavior), we suggest to approve the final air intake system in a climatic wind tunnel or on a natural testing area.

5 Conclusion

In cold and snowy regions snow particles, which are rolled up by vehicles, can cause an engine power drop. This paper shows possibilities to evaluate the risk of such an engine power drop. These possibilities include measurements on a natural test area, measurements in a climatic wind tunnel and numerical simulations. The measurements allow the evaluation of the snow amount, which penetrates into the air filter body, and the evaluation of the engine power drop. However, these tests are only possible in a later stage in the vehicle development process. If adjustments to the vehicle are required in this late stage, high costs could emerge. For this reason we suggest to introduce the described simulation method in the virtual testing stage. The simulation method allows an approximate calculation of the snow amount, which penetrates into the air filter body. If necessary, the air intake system can be improved at an early stage. Since there are simplifications applied in the simulation model (e.g., the wall-impact behavior), we suggest to approve the final air intake system in a climatic wind tunnel or on a natural testing area.

References

1. Huber, C., Weigand, B., Reister, H., Binner, T.: Use of an Eulerian/Lagrangian framework to improve the air intake system of an automobile with respect to snow ingress. SAE Int. J. Passeng. Cars Mech. Syst. **10**(2), 487–497 (2017)

2. Nakaya, U.: Snow Crystals: Natural and Artificial. Harvard University Press, Cambridge (1954)
3. Libbrecht, K.: Ken Libbrecht's Field Guide to Snowflakes. Voyageur Press, Minneapolis (2006). ISBN 978-0760326459
4. Auer Jr., A.H., Veal, D.L.: The dimension of ice crystals in natural clouds. J. Atmos. Sci. **27**, 919–926 (1970)
5. Heymsfield, A.: Ice crystal terminal velocities. J. Atmos. Sci. **29**(7), 1348–1357 (1972)
6. Ono, A.: Growth mode of ice crystals in natural clouds. J. Atmos. Sci. **27**(4), 649–658 (1970)
7. Huber, C., Weigand, B., Reister, H., Binner, T.: Modeling and numerical calculation of snow particles entering the air intake of an automobile. SAE Int. J. Passeng. Cars Mech. Syst. **8**(2), 538–545 (2015)
8. User guide STAR-CCM+, CD-adapco
9. Schiller, L., Naumann, A.: Über die grundlegenden Berechnungen bei der Schwerkraftaufbereitung. Z. Ver. Deut. Ing. **77**, 318–320 (1933)
10. Loth, E.: Drag of non-spherical solid particles of regular and irregular shape. Powder Technol. **182**(3), 342–353 (2008)
11. Sommerfeld, M.: Modelling of particle-wall collisions in confined gas-particle flows. Int. J. Multiph. Flow **18**(6), 905–926 (1992)
12. Higa, M., Arakawa, M., Maeno, N.: Measurements of restitution coefficients of ice at low temperatures. Planet. Space Sci. **44**, 917–925 (1996)
13. Higa, M., Arakawa, M., Maeno, N.: Size dependence of restitution coefficients of ice in relation to collision strength. Icarus **133**(2), 310–320 (1998)
14. Hauk, T., Roisman, I., Tropea, C.: Investigation of the impact behaviour of ice particles. In: Presented at the 6th AIAA Atmospheric and Space Environments Conference, USA, June 16–20 (2014)
15. Matsumoto, S., Saito, S.: Monte Carlo simulation of horizontal pneumatic conveying based on the rough wall model. J. Chem. Eng. Japan **3**, 223–230 (1970)
16. Crowe, C., Sommerfeld, M., Tsuji, Y.: Multiphase Flows with Droplets and Particles. CRC Press, Boca Raton (1998). ISBN 978-0849394690
17. Barnes, P., Tabor, D., Walker, J.C.F.: The friction and creep of polycrystalline ice. Proc. R. Soc. Lond. **324**, 127–155 (1971)
18. Villedieu, P., Trontin, P., Chauvin, R.: Glaciated and mixed-phase ice accretion modeling using ONERA 2D icing suite. In: Presented at the 6th AIAA Atmospheric and Space Environments Conference, USA, June 16–20 (2014)

Cooling Air Flow

Introduction of a New Full-Scale Open Cooling Version of the DrivAer Generic Car Model

Burkhard Hupertz[1(✉)], Lothar Krüger[1], Karel Chalupa[1],
Neil Lewington[2], Brendan Luneman[2], Pedro Costa[3], Timo Kuthada[4],
and Christopher Collin[5]

[1] Ford Werke GmbH, Spessartstr., 50725 Cologne, Germany
{bhupertz,lkruegel,kchalupa}@ford.com
[2] Ford Asia Pacific, 1735 Sydney Road, Campbellfield, VIC, Australia
{nlewingl,bluneman}@ford.com
[3] Ford of South America, 2000 Av., Henry Ford, Camacari, Brazil
pcosta37@ford.com
[4] Forschungsinstitut für Kraftfahrwesen und Fahrzeugmotoren Stuttgart,
Stuttgart, Germany
[5] Institute of Aerodynamics and Fluid Mechanics,
Technische Universität München, Munich, Germany

Abstract. Since the introduction of the generic aerodynamic research model DrivAer, an increasing amount of aerodynamic research and aerodynamic CAE method development activities have been based on this simplified generic car body. Due to the OpenSource nature of the model it has not only been used by academia but also by several automotive OEMs and CAE software developers. The DrivAer model has delivered high quality experimental data to permit validation of existing aerodynamic CAE capabilities and to accelerate the development of new more sophisticated numerical methods.

Vehicle aerodynamic performance is significantly influenced by the airflow through the engine bay. The current, closed cooling version of the DrivAer does not enable an assessment of the influence of the cooling airflow on the vehicle's aerodynamic characteristics.

A new open cooling version of the DrivAer model is proposed to further expand the usability of the overall DrivAer concept. Beyond an extended usage in vehicle aerodynamics the layout of the model will allow for investigations related to powertrain cooling, heat protection, brake cooling and wind noise.

This paper focuses on the conceptual layout of the Open Cooling DrivAer model and will explain the instrumentation concept of the physical test model. Furthermore initial wind tunnel test data of the baseline configuration will be presented.

1 Introduction

Vehicle aerodynamics plays an important role for multiple aspects of the vehicle development and is therefore a key area of active research. Beside the influence on fuel consumption and CO2 emissions, vehicle aerodynamics is of critical importance to engine and battery cooling, heat protection, brake cooling, soiling and aero-acoustics. Several generic car models, such as the Ahmed body [1], the SAE model [2] and the

J. Wiedemann (ed.), *Progress in Vehicle Aerodynamics and Thermal Management*,
https://doi.org/10.1007/978-3-319-67822-1_3

Windsor model [3] have been deployed to research aerodynamic characteristics of ground vehicles. A comprehensive overview of the different available models was presented by Le Good and Garry in 2004 [4].

Taking the value of generic car models for aerodynamic research to a next level, the Institute of Aerodynamics and Fluid Mechanics of the Technische Universität München (TUM) in cooperation with Audi AG and BMW group in 2012 introduced a new generic aerodynamic reference model called the DrivAer [5]. Derived from an Audi A4 and BMW 3 Series the DrivAer model represents a realistic vehicle design and includes car features like wheels and wheel-houses, a detailed underbody and mirrors, which have not previously been included in the generic research models.

The OpenSource nature of the DrivAer model and its relevance to automotive aerodynamics has given rise to a variety of research projects by various automotive OEMs, research institutes and CFD software vendors.

As the originator of the geometry TU Munich started using the model for aerodynamic investigations. These included the transient flow effects of the model under yawing conditions [6, 7]. Further work included the commissioning of the new rolling road system in their model scale wind tunnel [8], a capability study of an OpenSource CFD Software for unsteady vehicle aerodynamics [9] and work on wind tunnel test section layout and correction methods [10].

The Technical University Berlin was an early adopter of the concept building a quarter scale model in 2012 [11]. An analysis of the basic flow characteristics by Strangfeld et al. [12] highlighted the presence of a separation bubble at the rear window that is caused by geometrical constraints and an adverse pressure gradient. The investigations identified an unsymmetrical wake flow on the model. Further investigations [13] were conducted on active flow control with fluidic oscillators to influence this separation bubble.

Stoll et al. [14–16] used the model to investigate the transient aerodynamic effects on drag and side force in gusty wind conditions. It could be shown that the sensitivity of the performance of aerodynamic devices (e.g. trunk spoiler or wheel deflectors) in gusty wind conditions is different when compared to steady state measurements. Additional CFD investigations showed good correlation of the resulting forces.

The model is also well suited for CFD validation and method development. Guilmineau [17] used the model to investigate different simulation and meshing schemes for the NUMECA CFD solver. Jakarlic et al. (Technical University Darmstadt) validated a new eddy-resolving VLES model [18] by studying the influence of the underbody geometry and the wheel rotation on the aerodynamics of the DrivAer model [19]. Jungmann et al. [20] used the DrivAer model to validate selected numerical solution methods including unsteady Reynolds-averaged Navier-Stokes (RANS), an Improved Delayed Detached Eddy Simulation (IDDES) and a Lattice Boltzmann (LBM). Soares et al. [21] investigated the models wake development resulting from the three different rear-body configurations in the far-field using a RANS Methods. Simmonds et al. [22] presented a comprehensive study on three automotive models, two production type vehicles and the DrivAer notchback, with different cooling drag deltas using two commercial CFD solvers, one based on RANS and one based on LBM. Drag and lift coefficient values were compared and analyzed. Haag et al. [23] used the DrivAer model to validate a sliding mesh approach to represent different rotating rim types.

Beyond this, the model has already been utilized for the investigations of add-on parts like spoilers. LeGood et al. [24] introduced systematic add-on parts like a rear trunk spoiler, trunk lid extensions and front chin lip spoilers with different sizes. They showed the influence of these parts on DrivAer's aerodynamic characteristics and proposed a publicly available database for such add-on parts.

Ford has built three 1:1 scale physical DrivAer models which have been used for several CAE method development activities including an in depth yaw prediction investigation (not yet published). Besides supporting CAE method development, the physical models have been used for several aerodynamic studies including an extended correlation of the Ford aerodynamic wind tunnels in Dearborn and Merkenich and the Monash wind tunnel in Melbourne.

Building on the success of the DrivAer concept and further extending the relevance and usability of the generic model, Ford in cooperation with Institute of Aerodynamics and Fluid Mechanics of the Technische Universität München (TUM) and the Forschungsinstitut für Kraftfahrwesen und Fahrzeugmotoren Stuttgart (FKFS) have developed an open cooling (OC) version of the DrivAer.

The main purpose of this paper is to present the Open Cooling DrivAer (OCDA) to a broader audience and with that to lay the foundation for wider usage. The paper starts with a detailed description of the baseline OC DrivAer model, followed by a review of the concepts for model modularity and instrumentation. Subsequently, the physical models, the Ford 1:1 scale, the FKFS 1:4 scale and the TUM 1:2.5 scale model, will be reviewed. Initial wind tunnel test results of the Ford 1:1 scale OCDA model will be presented, followed by a comparison of the aerodynamic behaviour of the Ford full-size Open and Closed Cooling DrivAer models to the scale FKFS and TUM OC DrivAer models. Finally a proposal regarding an efficient sharing of OC DrivAer related wind tunnel test and CFD data will be discussed.

2 Development of the Open Cooling DrivAer Model

2.1 Background

The airflow through the engine bay of a vehicle has a significant influence on its overall aerodynamic performance and typically makes up 5–10% of the overall aerodynamic drag. Minimising the cooling drag is therefore an important factor in delivering a low-drag vehicle. Unfortunately, the complex interactions between the cooling airflow and the external airflow are not sufficiently understood. In particular, the ability of numerical methods to adequately predict the cooling airflow related drag change remains a challenge. The OC DrivAer is expected to facilitate studies to determine the mechanisms that influence cooling drag, without the geometrical and manufacturing uncertainties typically encountered when studying typical production vehicles.

While several generic vehicle models have been used for external aerodynamic research, the number of generic and publicly available models enabling specific aerodynamic investigations related to powertrain cooling, heat protection, brake cooling and aero acoustic is much more limited. A modified Ahmed Body, with ducted internal flows was presented by Barnard [25] with the aim to analyze the drag related to cooling air flows. Kuthada et al. [26, 27] introduced a SAE Model with internal flows based on the proposal of Cogotti [28] which includes rotating wheels and several options for the ducted internal flows. This geometric setup was also adopted by Baeder et al. [29] for the validation of underhood CFD simulations. In a later work, this model was extended to be used for validation work on heat transfer in the underfloor region [30], including a heated plug representing the exhaust system.

With the integration of a reasonably realistic engine bay into the DrivAer model Ford saw an opportunity to further increase the usability of the DrivAer beyond classical vehicle aerodynamics by extending the modularity of the model. Whilst the original DrivAer was limited to exchangeable tophats and underbodies, the OC DrivAer permits modifications related to but not limited to the front grilles, the front fascia, the upstream ducting, the cooling pack, the engine block, the engine undershield, the front wheel-houses and the A-pillar.

2.2 The Baseline Open Cooling DrivAer Model

The geometry of the Open Cooling (OC) DrivAer (Fig. 1) is based on the Closed Cooling (CC) DrivAer model with the detailed underbody proposed by Heft et al. in 2012 [5]. Like the CC DrivAer, the OC DrivAer also features three different body styles, a Fastback, Notchback and Estate.

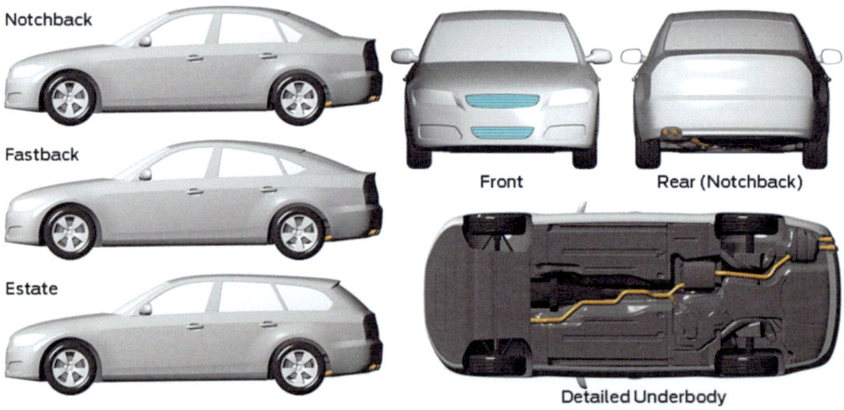

Fig. 1. Exterior geometry of the Open Cooling DrivAer

As shown in Fig. 2, the original CC DrivAer model has been complemented with a simplified, generic engine bay. To facilitate a well-defined underhood airflow the complete flow domain between the new designed front grilles, where the air enters the engine bay, and the prescribed air outlets into the wheel houses or the exhaust tunnel is 100% ducted. The airtight ducting consists of two parts, the ducting upstream of the cooling package (upstream ducting) and the full engine bay encapsulation.

The OC DrivAer model engine bay includes a cooling package and a powertrain with an attached exhaust system. Following the logic of the DrivAer concept, all components have been simplified as much as possible without compromising the

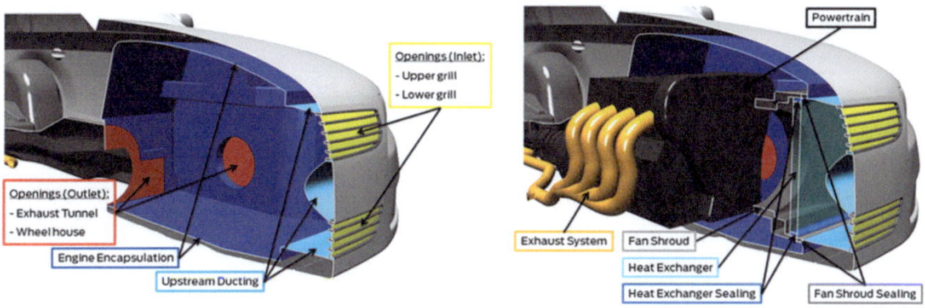

Fig. 2. Underhood package of the Open Cooling DrivAer

relevance of the component with respect to their impact on the engine bay airflow. As shown in Fig. 2, the cooling package features a simplified fan shroud and a single heat exchanger. The cooling package of the baseline OC DrivAer model does not include a cooling fan. The cooling package is completely sealed to the front face of the engine bay encapsulation so that all of the air flow entering the engine bay through the front grilles must flow through the heat exchanger and the downstream fan shroud. The longitudinal powertrain consists of an engine block and a transmission which partially sticks out of the engine bay into the underbody exhaust tunnel. On the right side of the engine, a 4-port exhaust manifold is attached to the powertrain. The exhaust manifold is connected to the exhaust system and is an extension to the design released on the CC DrivAer model.

To ensure that the open cooling DrivAer model is fully defined, the authors propose a pressure drop characteristic of the heat exchanger of the baseline model (OCDA HX1). The pressure drop is shown in Fig. 3 and reflects the behavior of a Ford production radiator instrumented with 60 pressure probes (s. Sect. 2.4).

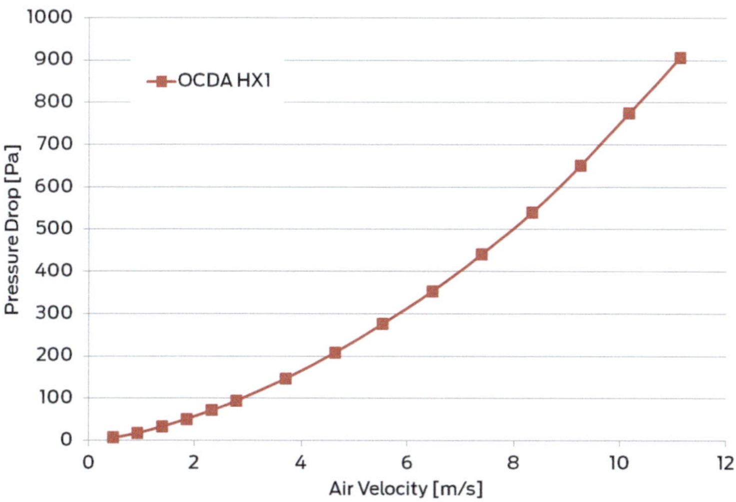

Fig. 3. Heat exchanger pressure drop of the Ford OCDA

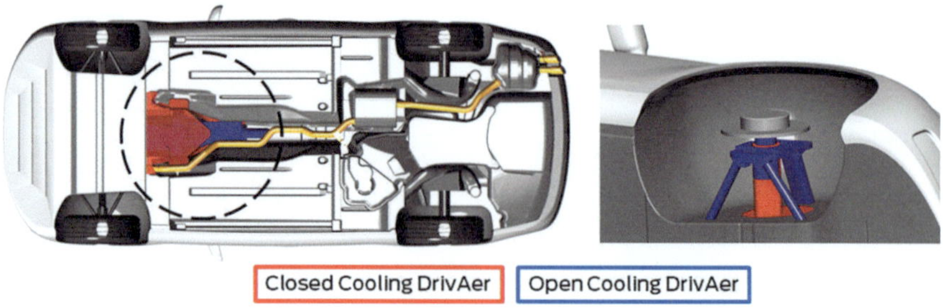

Fig. 4. Geometric differences (transmission, wheel mounting) between the new open cooling and original closed cooling DrivAer

While it was the goal of the authors to ensure that the OC DrivAer model remains identical to the CC version, it has to be recognized that the generic transmission housing included in the detailed underbody of the CC DrivAer has been replaced by a more realistic transmission in the OC DrivAer model (Fig. 4, left). Furthermore the introduction of openings in the wheel-houses resulted in a redesign of the wheel mounting concept, shown in Fig. 4, right. These differences must be considered when comparing test or CFD results of the CC and a blanked version of the OC DrivAer.

Following the OpenSource concept which laid the foundation for the success of the original DrivAer model, the CAD and the heat exchanger pressure drop data of the OC DrivAer is available in the public domain in ANSA, STEP and STL format via the following web sites: www.drivaer.com or www.ecara.org.

2.3 Variability of the Ford Open Cooling DrivAer Model – Beyond Aerodynamics

The original DrivAer model already offers configuration flexibility via an exchangeable vehicle rear end (notchback, fastback and estate) and underbody (flat and detailed) extending the usefulness of the generic model. With the introduction of the open cooling DrivAer the opportunity arises to further extend the modularity and with that the usability of the model.

Based on the presented baseline OC DrivAer model, Ford has developed a highly modular concept that enables additional aerodynamic investigations but also permits studies which go beyond classical aerodynamics.

As shown in Fig. 5, the upper and lower front grilles can be exchanged as well as the complete front fascia. With that different front end opening concepts and their impact on aerodynamics and cooling airflow can be assessed, such as different opening size, grille design and location. Furthermore the concept has been laid out such that air curtain or brake cooling ducts can be integrated in a revised front fascia.

Inside the engine bay, an exchangeable upstream ducting will support the integration of a grille shutter system. The cooling module can be removed, permitting assessments of differing pressure losses or multiple heat exchanger configurations. The model is equipped with an interchangeable fan shroud so that specific assessments of leakage flow, powered fan units, dual fans and speed-flaps are permissible. Finally, the

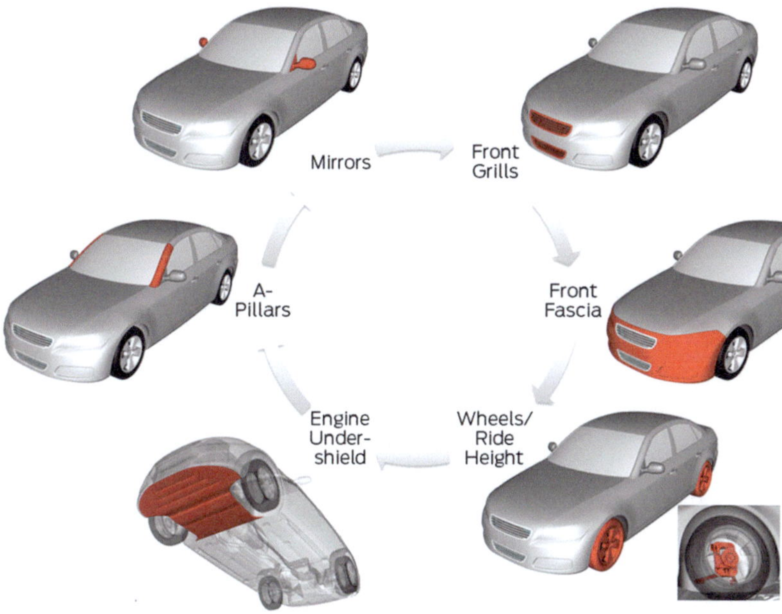

Fig. 5. Variability of the Ford Open Cooling DrivAer model

x-position of the powertrain can be modified to influence the back pressure on the cooling pack.

An interchangeable engine undershield will provide the opportunity to study the impact of air evacuation concepts on the cooling airflow and vehicle aerodynamics. The integration of NACA ducts into the engine undershield will support studies related to the convective cooling of components inside the engine bay. Furthermore brake cooling ramps or front wheel air deflector concepts can be investigated by modifying the engine undershield. Finally the mirrors and the a-pillar of the OC DrivAer model can be replaced enabling aerodynamic and exterior noise source related studies.

2.4 Instrumentation of the Ford Open Cooling DrivAer Model

To ensure the usefulness of the OC DrivAer, for in-depth aerodynamics and cooling airflow related studies, it is of critical importance to enable detailed measurements of relevant flow properties. The 1:1 scale Ford OC DrivAer models (notchback, fastback, estate) are equipped with 295 pressure probes (Fig. 6). The models include 108 of the 114 probes suggested for the CC DrivAer [5] plus an additional 75 probes on the upperbody and underbody to define detailed pressure maps on the side window, the C-pillar, the vehicle rear end and the underbody region.

The outflow of cooling air from the engine bay into the wheel-house is expected to have a significant impact on the overall aerodynamic performance of a vehicle. Wheel well pressures will be monitored via 26 pressure probes in the front left wheel-house. Engine bay pressures, specifically in the upstream ducting, on the fan shroud, on the

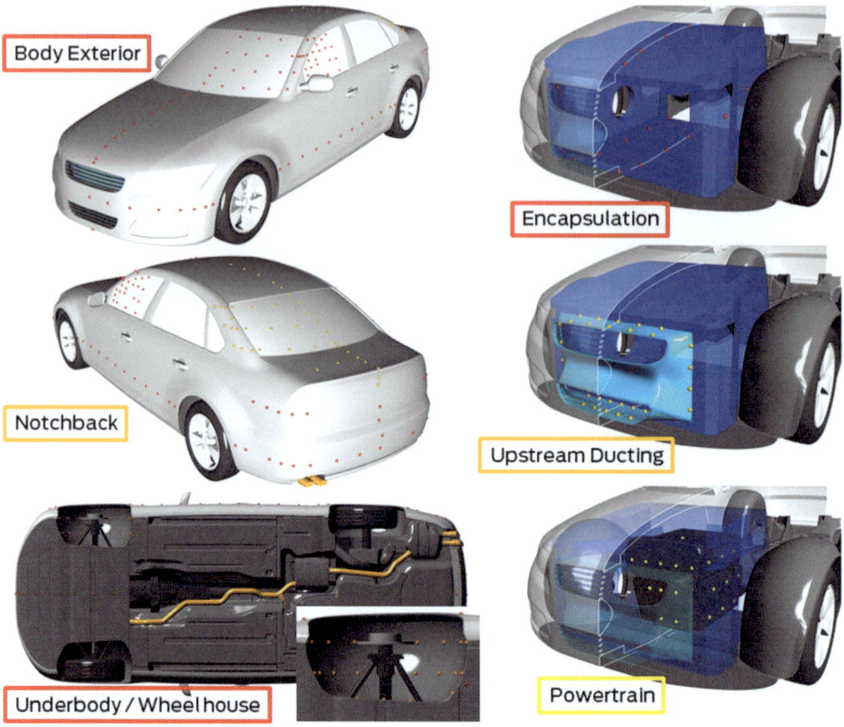

Fig. 6. Pressure tab distribution of the Ford Open Cooling DrivAer model

engine bay encapsulation and on the powertrain, are monitored via 86 pressure probes. These probes are expected to deliver valuable information in particular when investigating cooling package changes, air evacuation concepts or NACA ducts. All pressure probe locations including the proposed numbering are included in the publically available CAD data.

Monitoring cooling airflow is a key objective for the OC DrivAer and as such it is essential to be able to accurately determine the mass flow rate through the engine bay and the cooling package. The Ford OCDA model incorporates an instrumented radiator with 60 equally distributed FKFS radiator probes. Careful routing of the pressure tubes on the instrumented radiator (s. Fig. 7), ensures an instrumentation blockage ratio below 1%. After calibration, this method can determine the volume flow rate with a maximum deviation of less than 3%. An advantage of this method is that the radiator stays functional and measurements are also possible with a running fan. Measurements take approx. 30 s and can be obtained simultaneously to aerody-namic force measurements in the wind tunnel [35].

Finally, the 1:1 scale Ford OC DrivAer model features an optical access to the engine bay. Large acrylic glass windows have been integrated in the hood and the upper shell of the engine bay encapsulation (s. Sect. 3.1). The optical access will enable flow field measurements inside the engine bay using Particle Image Velocimetry (PIV) or Laser-Doppler Anemometry (LDA).

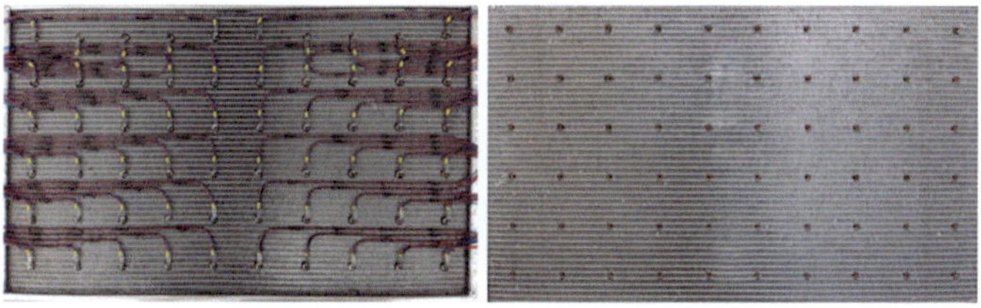

Fig. 7. Open Cooling DrivAer heat exchanger instrumented with 60 pressure probes (left: rear view, right: front view)

3 Physical Models

To the best knowledge of the authors so far 3 physical models of the open cooling DrivAer have been built and will be described in the following sections: The Ford 1:1 scale, the FKFS 1:4 scale and the TUM 1:2.5 scale model.

3.1 The Ford 1:1 Scale Open Cooling DrivAer Model

The 1:1 scale model built by Ford (s. Fig. 8) is 100% representative of the CAD model described in the previous chapter. Three different rear ends, originally manufactured for the closed cooling DrivAer model can be used with the new open cooling version. During the design and build process, special attention has been paid to the airtightness of the complete underhood flow domain, as unwanted air leakage, e.g. due to the deflection of parts under wind load, can have a significant impact on the aerodynamic test results.

To validate the geometric integrity of the physical model the final product has been scanned. The geometrical deviation of the physical model to the CAD model is shown

Fig. 8. The Ford 1:1 scale notchback Open Cooling DrivAer Model

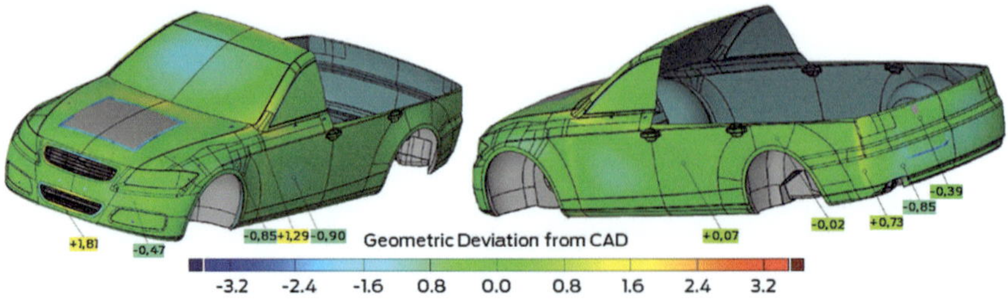

Fig. 9. Geometric deviation of the Ford OC DrivAer model from CAD

in Fig. 9. In the majority, the 1:1 scale model does not deviate more than ±1 mm from the original CAD. A maximum deviation in localised areas remains less than ±2 mm.

As outlined in the previous chapter Ford has developed a highly modular OC DrivAer concept. The physical model permits a suite of geometric modifications: front grilles, front fascia, heat exchanger, fan shroud, engine under shield, wheels, mirrors and A-pillar. Figure 10 shows some examples of the modularity of the Ford Open Cooling DrivAer model. Furthermore the acrylic glass windows integrated in the hood and the engine encapsulation can be seen in Fig. 10.

Fig. 10. Examples of the modularity of the Ford Open Cooling DrivAer model Left: Exchangeable upper and lower front grill. Middle: Exchangeable A-pillar. Right: Acrylic glass windows in the hood and engine encapsulation. Movability of the powertrain and the fan shroud.

The OC DrivAer model is equipped with solid tires which match the CAD model (s. Fig. 11). Due to the rigid tires uncontrolled tire deformation, which is known to have a significant impact on the aerodynamic behavior of a vehicle, has been avoided.

Unfortunately the rigid tires cannot be used for moving ground tests due to expected model vibrations during the test. To enable moving ground testing, a second set of productions wheels, similar to the DrivAer CAD design are implemented (s. Fig. 11). These wheels have been scanned and a CAD model of the wheels has been reengineered from the scan.

Fig. 11. Wheels of the 1:1 scale Ford Open Cooling DrivAer model

3.2 The TUM 1:2.5 Scale Open Cooling DrivAer Model

The 40% scale model of the TUM was introduced by Heft et al. [5] in 2012, including a proposal for a geometry for cooling air flow investigations. Since then the model has been in use by the TUM for a variety of research projects. The standard configuration for wind tunnel testing with moving ground simulation in the large wind tunnel of the TUM is presented in Fig. 12. The experimental setup includes a model support via a top strut and four wheel struts. Consequently, no connection between the model and wheels is present [8].

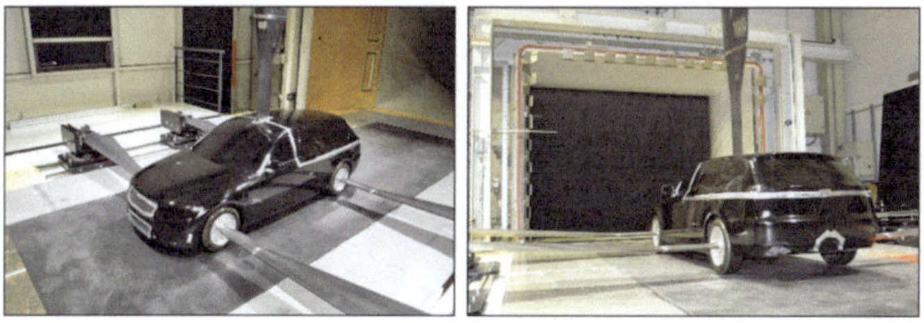

Fig. 12. 40% scale DrivAer model inside the open jet test section of large TUM wind tunnel.

In 2016, the existing model was updated with the latest underhood geometry, which offered the opportunity of a more realistic engine bay geometry. The main focus was put on a thorough interior sealing in order to minimize the effect of leakage air flow. Therefore, both the inlet duct as well as the radiator package were manufactured from milled aluminum (see Fig. 13, left).

The entire assembly is built as a stack of components – namely inlet duct, honeycomb 1, perforated sheet, honeycomb 2 and fan shroud, see Fig. 13, right. The interfaces between adjacent parts are realized with sealing elements. The stack-wise structure offers an adequate flexibility towards adjustments of the radiator characteristics.

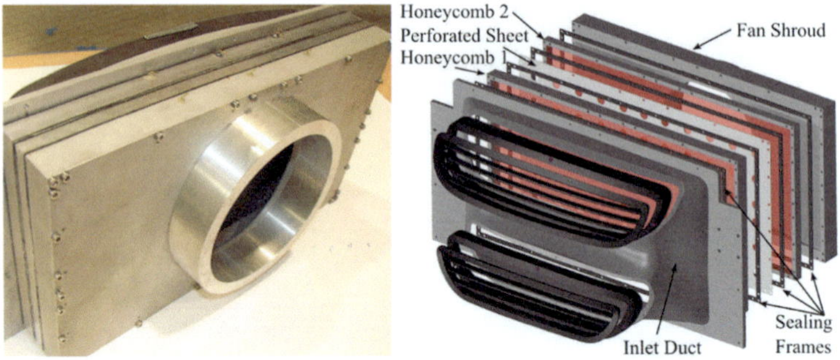

Fig. 13. Inlet duct and radiator package from milled aluminum (left). Overview of particular components (right).

The inlet duct is mounted to the longitudinal beams of the wind tunnel model using fixing brackets, see Fig. 14, left. All potential gaps between inlet duct, upper and lower grille and to the outer shell of the model are closed with sealing compound. With this approach the radiator package, which is mounted to the inlet duct, can be accessed and adjusted from the bottom of the model without removing the model from the wind tunnel test section.

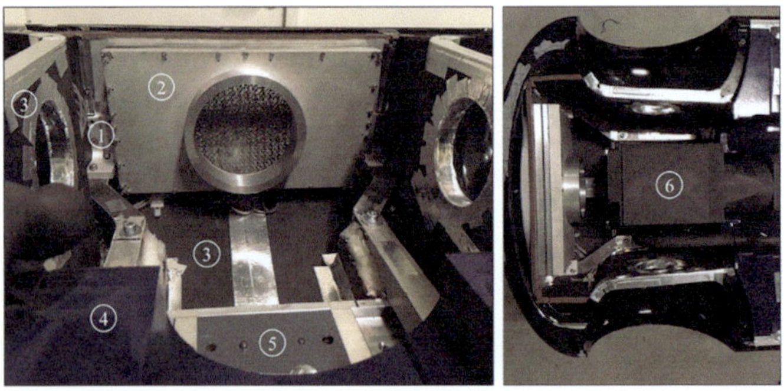

Fig. 14. Engine bay design of the 40% scale model of the TUM with (right) and without engine and gearbox (left). The engine undershield was removed for the picture (1: fixing brackets, 2: radiator package, 3: engine compartment panels, 4: underfloor inset, 6: engine and gearbox).

The parts such as engine bay compartment panels, engine, gearbox, exhaust pipes, upper and lower grille as well as the engine undershield are manufactured from rapid prototyping based on the selective laser sintering (SLS) technology. The underfloor inset was produced as a milling part from rigid foam. The complete assembly of the underfloor geometry is presented in Fig. 15. The model was upgraded with 28 pressure taps on the surface of the engine as well as three pressure tap's in front and behind the flow straighteners of the radiator.

Fig. 15. Front (left) and rear view (right) of the underfloor geometry.

3.3 The FKFS 1:4 Scale Open Cooling DrivAer Model

In 2014 FKFS built two 25% scale models of the DrivAer with an underhood compartment [32]. These models are used during courses with students and for basic research [33]. Compared to the results presented earlier [34], the models had to be modified to meet the changed geometry specifications. Major changes have been implemented in the front wheel arches (wheel house openings and suspensions, see Fig. 16, right) and the ducting of the radiator and the transmission tunnel (see Fig. 16, left)

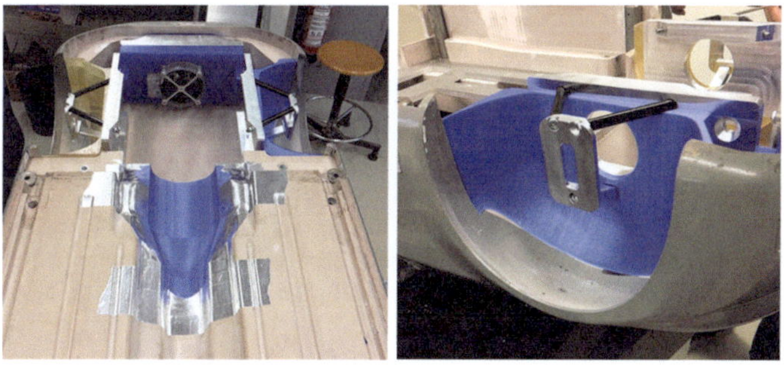

Fig. 16. Geometry changes in the underbody/underhood (left) and wheel arches (right) of the 25% scale DrivAer models at FKFS

4 Experimental Results

The following section will provide an overview of 1:1 scale wind tunnel test results of the full scale Ford Open Cooling DrivAer model. Test data presented in this paper refers to wind tunnel tests conducted in the Pininfarina wind tunnel under static floor conditions (Ground Effect Simulation System (GESS) off). In some explicitly mentioned cases test data was obtained under moving ground conditions (GESS on). A comparison of the measured aerodynamic coefficients of the new open cooling and the original closed cooling DrivAer model and a comparison of the aerodynamic coefficients of the 3 different open cooling DrivAer models presented in Sect. 3 is included.

4.1 Ford 1:1 Scale Open Cooling DrivAer

Figure 17 shows an overview of the aerodynamic force of the baseline OCDA model and several design variants/test configurations. Excluding the last configuration (Prod. Wheels/GESS on) all configurations are tested under static floor conditions (GESS off). The last configuration uses the Pininfarina T-Belt system in combination with a boundary layer treatment via a scoop and a tangential blowing system [36].

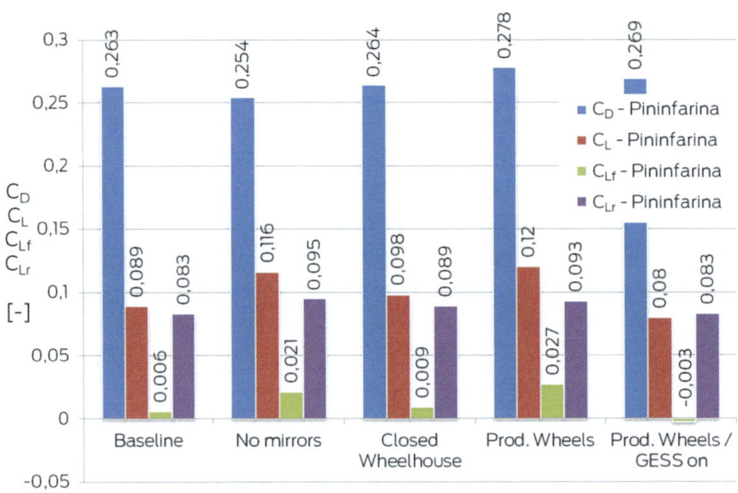

Fig. 17. Overview of the aerodynamic forces of several OCDA variants/test configurations. Excluding the last configuration (Prod. Wheels/GESS ON) all configurations are tested under fixed floor/wheels conditions in the Pininfarina wind tunnel.

Starting with the aerodynamic drag, the data shows that the removal of the mirrors results in a 9 counts drag reduction ($\Delta C_D = 0.009$). Interestingly, blanking the openings from the engine bay into the left and right side wheel-house results in a slight drag increase. It was originally anticipated that closing the wheel-house would positively influence the overall drag via two mechanisms. Firstly, the wheel-house blanking should reduce the cooling airflow and with that reduce the cooling airflow drag and hence the overall drag. Secondly, eliminating the outflow from the engine bay into the wheel-house and thus eliminating the complex interactions between the cooling airflow and the external airflow around the front wheels was expected to lead to a reduced overall aerodynamic drag. Further investigations will be required to develop a better understanding of the controlling mechanisms. Replacing the OCDA solid wheels with production wheels (s. Fig. 11) results in a 15 counts drag increase. As expected, changing the test condition from static floor to full moving ground for the OCDA model with production tires results in a 9 count drag reduction. Front and rear lift measurements (Fig. 17) indicate that removing the mirrors lead to a slight increase in front and rear lift. Similar to the effect on C_D closing the wheel-house openings resulted only in a minor change in front and rear lift. The change to production wheels increased the lift in the front and the rear. Changing the test condition to moving ground, leads to an expected global reduction of the front and the rear lift.

Figure 18 depicts the pressure distribution of the Open Cooling DrivAer for the open cooling and closed cooling (blanked upper and lower grill openings) configuration. Closing the grille openings results in significant local pressure difference in the area of the leading edge of the hood and a small but consistent offset of approximately $\Delta Cp = 0.005$ in the rear of the vehicle. On the backside of the vehicle a base pressure difference of approximately $\Delta Cp = -0.01$ was measured. Pressure differences in the underbody region between open and closed grille configurations are more notable. In the front part of the underbody the pressure level observed in the closed cooling case is substantially lower than in the open cooling case. This is most likely due to the increased amount of airflow underneath the vehicle resulting from closing the front grille openings. Furthermore, the influence of the airflow entering the underbody region from the engine bay through the opening in the exhaust tunnel starting at approximately x = 400 mm is clearly visible.

As shown in Fig. 19, a relatively constant pressure is measured in the wheel-house, between $Cp = -0.08$ and $Cp = -0.18$, aside from measurements at probes 550 and 557 (the front and rear lower regions of the wheel-house). In the lower front area of the wheel-house the pressure is significantly lower while on the lower rear side of the

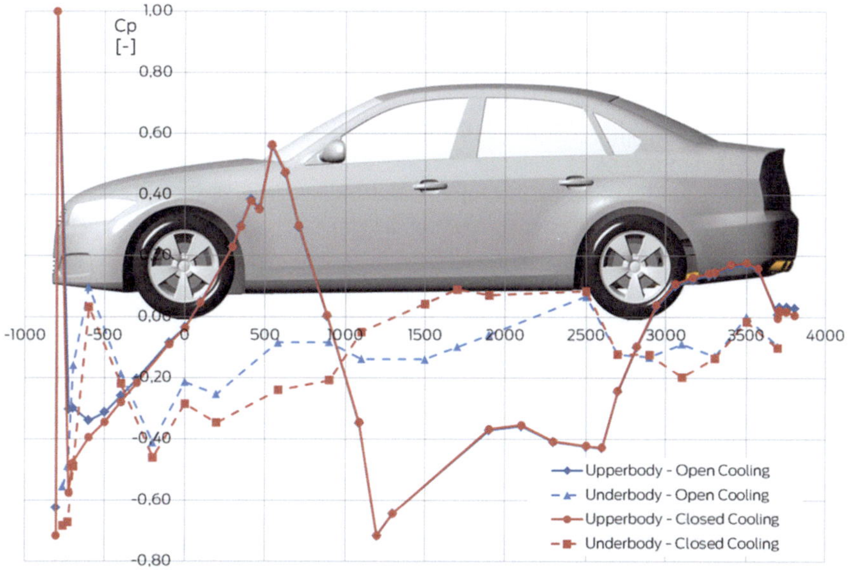

Fig. 18. Centerline pressure distribution (upper- and underbody) of the OCDA in the open cooling and closed cooling (blanked grill openings) configuration

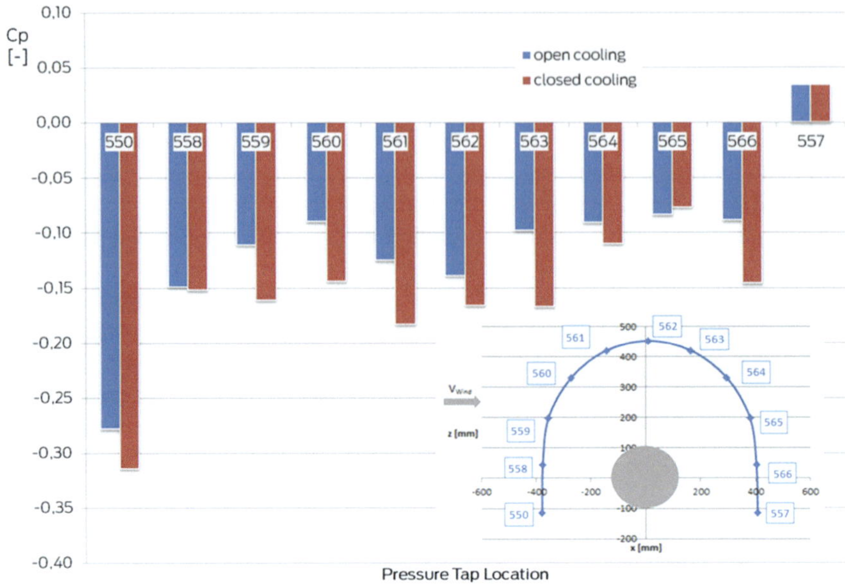

Fig. 19. OCDA pressure distribution (Cp) in the front left wheel-house at y = 750 mm in the open cooling and closed cooling (blanked grill openings) configuration

wheel-house a positive pressure level was measured. Opening the front grilles results in a global increase in wheel-house pressure, most likely due to the airflow entering the wheel-house through the engine bay.

4.2 Cooling Airflow Drag

As explained in the introduction the Open Cooling DrivAer model is expected to improve the understanding of the influence of the cooling airflow on the overall aerodynamic behavior of a vehicle. Figure 20 shows the cooling airflow drag ($\Delta C_{D\_Cooling}$) for the baseline OCDA model and several OCDA variants. The baseline model shows a relatively low cooling airflow drag of 10 counts which is expected to be a result of the perfect sealing of the underhood airflow domain. It is planned to conduct investigations of OCDA variants, which feature a non-perfect sealing of the fan shroud, to confirm this assumption. Removal of the mirrors has a negligible impact on $\Delta C_{D\_Cooling}$. Consistent to the drag values reported in Sect. 4.1, blanking the openings from the engine bay into the left and right side wheel-house results in a slight increase of the cooling airflow drag. Changing the wheels of the OCDA model (s. Fig. 11) results in a 4 counts larger cooling airflow drag which underlines the importance of the interaction between the cooling airflow and the exterior airflow. While changing the wheel design results in a measureable change in $\Delta C_{D\_Cooling}$, changing from a static floor test configuration to a full moving ground simulation in the Pininfarina wind tunnel only causes a 1 count change in $\Delta C_{D\_Cooling}$. More investigations are required to develop a better understanding of the wheel design and ground simulation impact on cooling airflow drag.

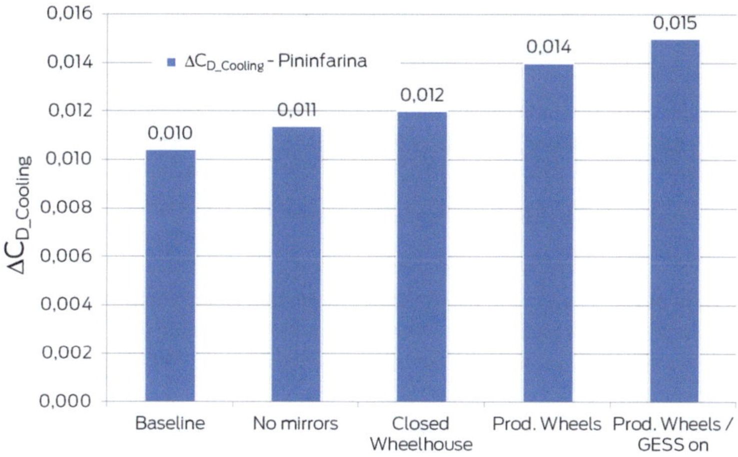

Fig. 20. Cooling airflow drag $\Delta C_{D\_Cooling}$ for different variants of the Open Cooling DrivAer model

4.3 Comparison to Original Closed Cooling DrivAer

As discussed in Sect. 2.2 there are a few geometrical alterations which might cause differences in the aerodynamic behavior between the original Closed Cooling DrivAer and the Open Cooling DrivAer model with blanked grille openings.

As shown in Fig. 21 there is almost no difference between the two vehicles in aerodynamic drag ($\Delta C_D = 0.001$) and rear lift ($\Delta C_{Lr} = 0.002$). A slightly higher

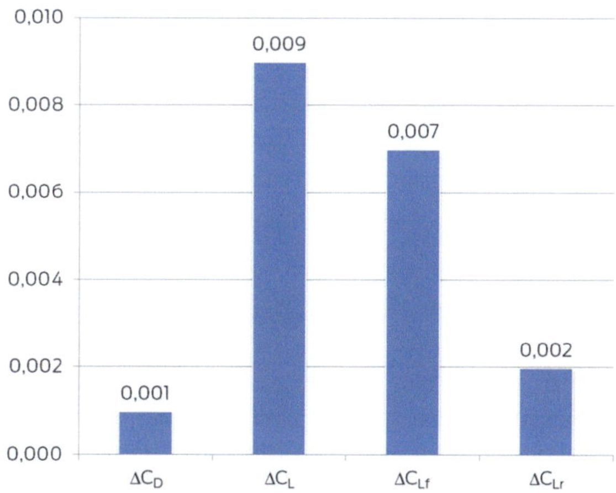

Fig. 21. Difference in aerodynamic forces between the Closed Cooling DrivAer and the Open Cooling DrivAer with blanked grille openings (Note: Data relates to vehicle configuration without mirrors)

difference of $\Delta C_{Lf} = 0.007$ has been measured for the front lift which results in a total lift difference of $\Delta C_L = 0.009$. Despite these differences it can be concluded that the overall aerodynamic behavior of both vehicle variants is almost identical. The documented similarity can serve as a foundation when comparing data, test and CFD, obtained for the two versions of the DrivAer model.

4.4 Comparison Between Scale and Full Scale Models

A comparison of the aerodynamic forces measured for the three different models in three different scales (Fig. 22) reveals differences in the absolute values. This is not unexpected and the reasons are multifaceted, e.g. small geometric differences between the models (s. Sect. 3) as well as the model support for a 5-belt system (Ford and FKFS) compared to a separated model and wheel support for the TUM single-belt system.

Furthermore, all reported force coefficients are not corrected for wind tunnel interferences effects and therefore differences resulting from the different wind tunnels have to be expected. Due to the differences in model scale, the results are presented for different velocities and, hence, Reynolds numbers (FKFS: *Re* = 3.5 Million, Ford: *Re* = 11.5 Million, TUM: *Re* = 5.2 Million), which can be an further explanation for the differences. With the availability in different scales, this model can also be a further test case for the investigation of wind tunnel interference effects and the further improvement of the respective correction methods.

However, the reason and focus for the development of OC DrivAer Model has been the underhood flow. When comparing the cooling air deltas, the models show the desired similarity. In cooling drag, the FKFS and the Ford model show a minor deviation of 0.001, whereas the TUM model is three counts lower compared to the Ford model. The different wheel mounting concept of the TUM model most likely causes the slightly different cooling drag. Looking at the effect of underhood flow on the lift coefficients, Ford and TUM measure the same results for both front and rear lift coefficients whereas the FKFS models shows a shift towards a larger effect on front lift rather than rear lift.

It also should be mentioned that the pressure drop of the radiator package of the two scale models slightly deviates from the reference curve provided by the Ford (see Fig. 23). The deviations remained in the range of $\Delta p < 10\%$.

5 Open Cooling DrivAer Data Sharing

To further improve the value of future Open Cooling DrivAer related investigations, the authors propose a standard format for the publication and exchange of Open Cooling DrivAer related wind tunnel test data and CFD results. Similar to the EADE aero data exchange two spreadsheet templates, one for wind tunnel test data (s. Fig. 24) and one for CFD results (s. Fig. 25), have been developed that summarize the key information

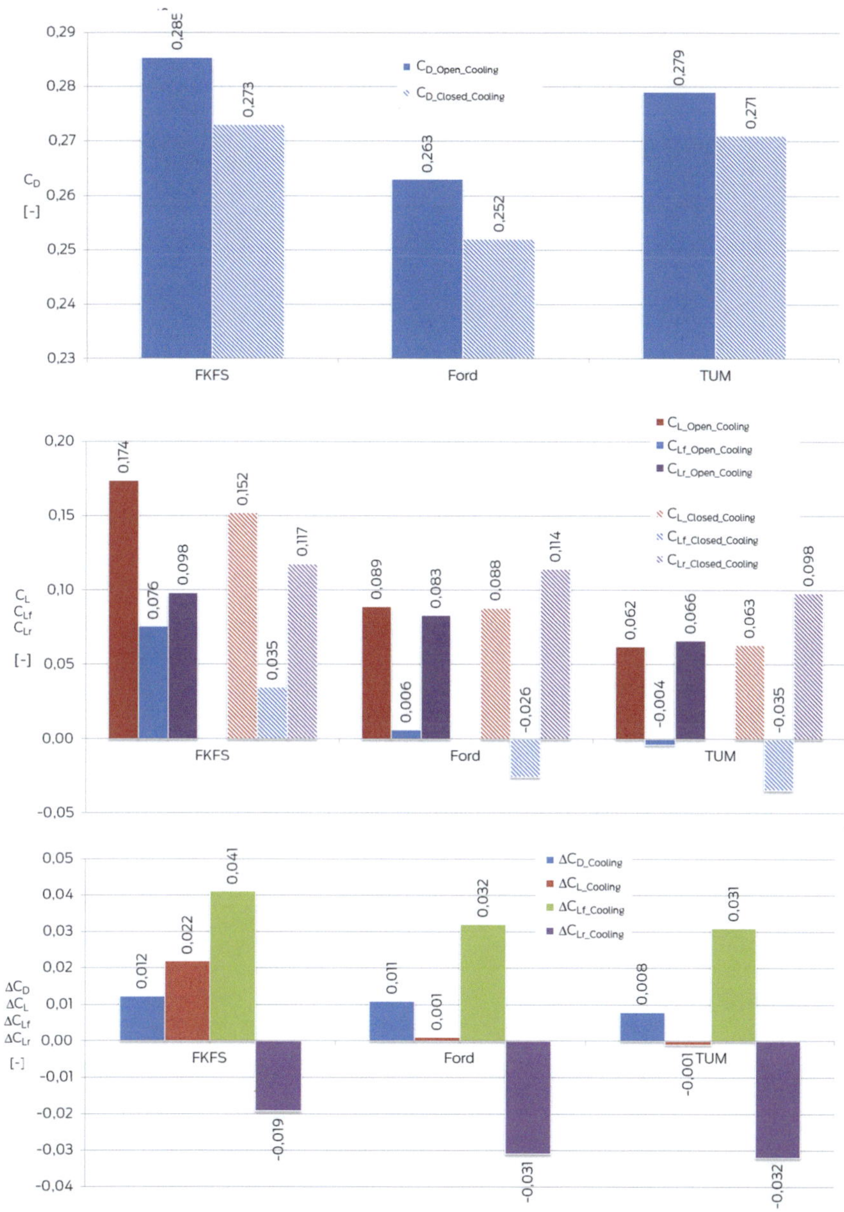

Fig. 22. Aerodynamic forces of the Ford 1:1 scale, the FKFS 1:4 scale and the TUM 1:2.5 scale model. (Top: Aerodynamic Drag C_D; Middle: Aerodynamic Lift: C_L, C_{Lf}, C_{Lr}; Bottom: Cooling airflow related force changes: $\Delta C_{D\_Cooling}$, $\Delta C_{L\_Cooling}$, $\Delta C_{Lf\_Cooling}$, $\Delta C_{Lr\_Cooling}$)

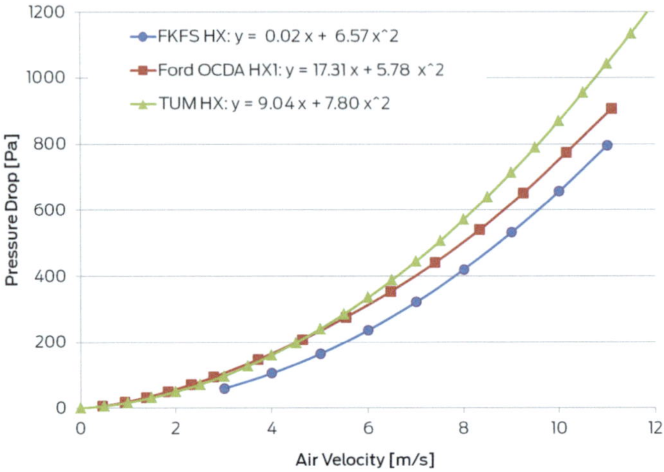

Fig. 23. Pressure drop for the radiator package of the Ford 1:1 scale, the FKFS 1:4 scale and the TUM 1:2.5 scale model.

required to make the data useful in the public domain. Both templates enable a precise identification of the investigated vehicle variant, which is a prerequisite for the usability of the shared data. Furthermore the templates provide information regarding the method, wind tunnel or CFD, employed to generate the data. Finally, the templates include a proposal for a selective result data sharing. Beside the aerodynamic forces the result proposal includes the cooling air mass flow, the upper- and underbody centerline pressure distribution and a reference pressure at a defined position in the engine bay and the left front wheel-house.

The templates as well as the test and CFD data provided via the templates will be stored and shared via the ECARA web-site: www.ecara.org

6 Summary

In this paper, an extension of the popular generic aerodynamic research vehicle, the DrivAer model, has been introduced. The integration of a generic underhood airflow domain into the original closed cooling DrivAer model further increases the usability and relevance of the model for aerodynamic research and CFD validation studies. To extend the DrivAer concept, a modular DrivAer model design has been proposed which will enable investigations related to powertrain cooling, heat protection, brake cooling and wind noise. To maximize the usefulness of the OCDA, a detailed instrumentation concept has been developed which is expected to deliver an in-depth understanding of the flow field of the generic vehicle model. A future oriented data sharing proposal included in the paper is expected to further increase the value of the Open Cooling DrivAer model.

The comparison of the aerodynamic forces of the original Closed Cooling and the Open Cooling DrivAer model showed very similar results. Initial measurements of the three Open Cooling DrivAer models, the Ford 1:1 scale, the FKFS 1:4 scale and the TUM 1:2.5 scale model, revealed differences in the aerodynamic forces, most likely resulting from differences in the test environment and setup. However the observed changes in aerodynamics forces resulting from the cooling airflow are very consistent and comparable results.

The presented wind tunnel test data obtained for the 1:1 scale Ford Open Cooling DrivAer model is expected to serve as a highly valuable foundation for an OCDA test and CFD database.

7 Future Work

Building on the modular and variable concept of the Open Cooling DrivAer model Ford will use the generic model for a variety of investigations including: (1) cooling airflow and cooling drag, (2) wheel aerodynamics, (3) air curtains and (4) aerodynamic and environmental wind tunnel correlation. In all cases the OCDA studies are expected not only to improve aerodynamic understanding but will provide an improved correlation dataset for CFD methods and tools.

Furthermore Ford, FKFS and TUM consider conducting further investigations to understand and minimize the observed differences in the aerodynamic forces of the different scale models.

Abbreviations

CC	Closed Cooling
CCDA	Closed Cooling DrivAer
GESS	Ground Effect Simulation System
IDDES	Improved Delayed Detached Eddy Simulation
LBM	Lattice Boltzmann Method
OC	Open Cooling
OCDA	Open Cooling DrivAer
RANS	Reynolds-averaged Navier-Stokes

Appendix

See Figs. 24 and 25.

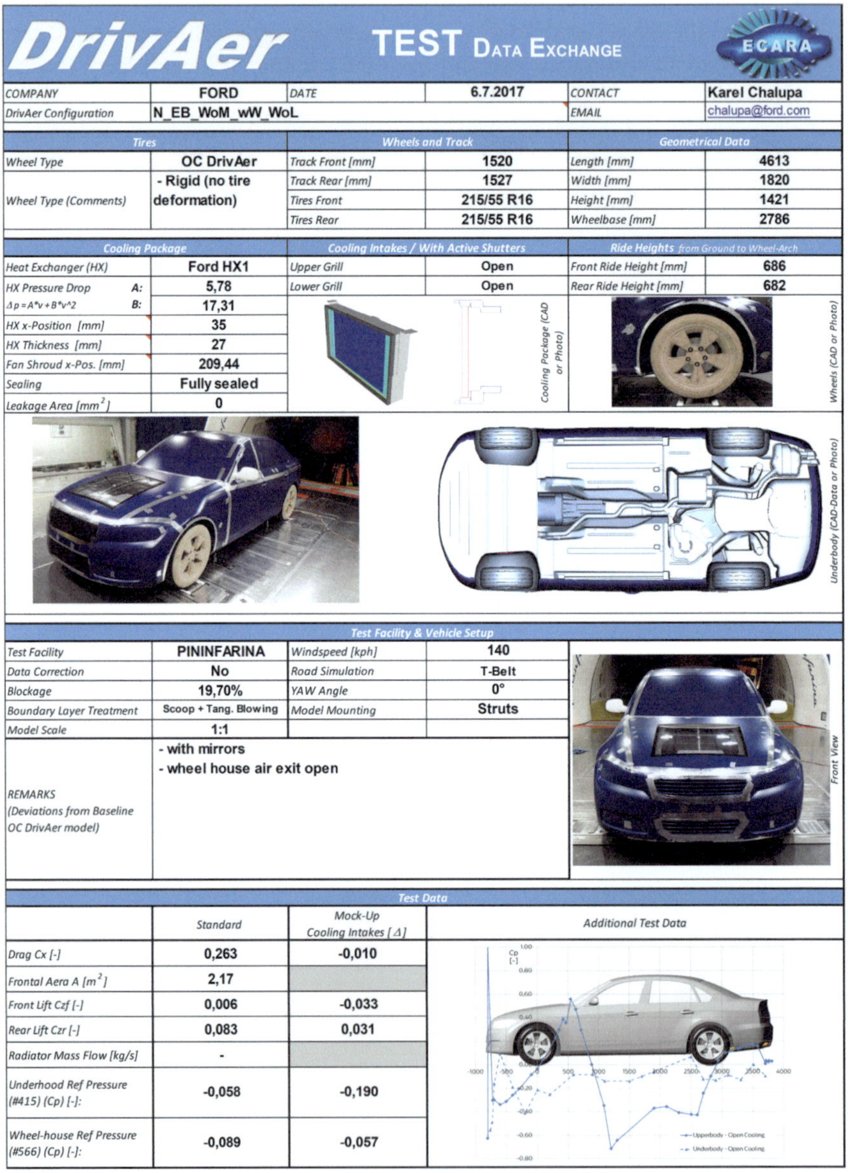

DrivAer TEST DATA EXCHANGE ECARA

COMPANY	FORD	DATE	6.7.2017	CONTACT	Karel Chalupa
DrivAer Configuration	N_EB_WoM_wW_WoL			EMAIL	chalupa@ford.com

Tires		Wheels and Track		Geometrical Data	
Wheel Type	OC DrivAer	Track Front [mm]	1520	Length [mm]	4613
Wheel Type (Comments)	- Rigid (no tire deformation)	Track Rear [mm]	1527	Width [mm]	1820
		Tires Front	215/55 R16	Height [mm]	1421
		Tires Rear	215/55 R16	Wheelbase [mm]	2786

Cooling Package		Cooling Intakes / With Active Shutters		Ride Heights from Ground to Wheel-Arch	
Heat Exchanger (HX)	Ford HX1	Upper Grill	Open	Front Ride Height [mm]	686
HX Pressure Drop A:	5,78	Lower Grill	Open	Rear Ride Height [mm]	682
Δp=A*v+B*v^2 B:	17,31				
HX x-Position [mm]	35				
HX Thickness [mm]	27				
Fan Shroud x-Pos. [mm]	209,44				
Sealing	Fully sealed				
Leakage Area [mm²]	0				

Cooling Package (CAD or Photo)

Wheels (CAD or Photo)

Underbody (CAD-Data or Photo)

Test Facility & Vehicle Setup			
Test Facility	PININFARINA	Windspeed [kph]	140
Data Correction	No	Road Simulation	T-Belt
Blockage	19,70%	YAW Angle	0°
Boundary Layer Treatment	Scoop + Tang. Blowing	Model Mounting	Struts
Model Scale	1:1		
REMARKS (Deviations from Baseline OC DrivAer model)	- with mirrors - wheel house air exit open		

Front View

Test Data	Standard	Mock-Up Cooling Intakes [Δ]
Drag Cx [-]	0,263	-0,010
Frontal Aera A [m²]	2,17	
Front Lift Czf [-]	0,006	-0,033
Rear Lift Czr [-]	0,083	0,031
Radiator Mass Flow [kg/s]	-	
Underhood Ref Pressure (#415) (Cp) [-]:	-0,058	-0,190
Wheel-house Ref Pressure (#566) (Cp) [-]:	-0,089	-0,057

Additional Test Data

Fig. 24. Open Cooling DrivAer wind tunnel test data reporting template

DrivAer CFD DATA EXCHANGE ECARA

COMPANY	FORD	DATE	09.05.2017	CONTACT	B Luneman
DrivAer Confiuration	N_EB_WoM_wW_WoL			EMAIL	bluneman@ford.com

CFD Solver		Mesh Settings		Tunnel Size	
Vendor	CD-adapco	Mesh Type	trim	WT length upstream [m]	20
Software	StarCCM+	Minimum Cell Size [mm]	2,5	WT length downstream [m]	40
Version	9.04.009	Total Number Cells [$x10^6$]		WT width / height [m]	22 / 9

Flow Boundary Conditions		Turbulence Settings		Vehicle Options	
Yaw Angle [deg]	0	Turbulence Model Class	DES	Bodystyle	Notchback
Vehicle Speed [ms-1]	38,889	Turbulence Model	k-w SST	Mirrors	Yes
Density [kg m-3]	1,18415	Near wall treatment	Hybrid	Front Ride Height (mm)	686
Absolute Ref. Pressure [Pa]	101325	Compressible Flow	No	Rear Ride Height (mm)	682
Solution Method	Transient - Avg.				

Heat Exchanger/Cooling Package Data		Cooling Configuration	Vehicle Configuration	
Heat Exchanger	Ford HX1	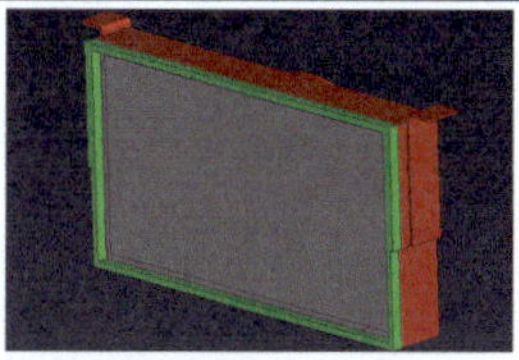 Cooling Package (CAD)	UPPER COOLING INTAKE	Open
HX Pressure Drop A:	5,78		LOWER COOLING INTAKE	Open
$\Delta p = A*v + B*v^2$ B:	17,31		Wheel Type	OC DrivAer
HX x-Position [mm]	35		Wheel Type (Comments)	- rigid, no deformation
HX Thickness [mm]	27		Road Simulation	No
Fan Shroud x-Pos. [mm]	209,44		Wheel Simulation	Static
Sealing	Fully Sealed			
Leakage Area (mm^2)	0			

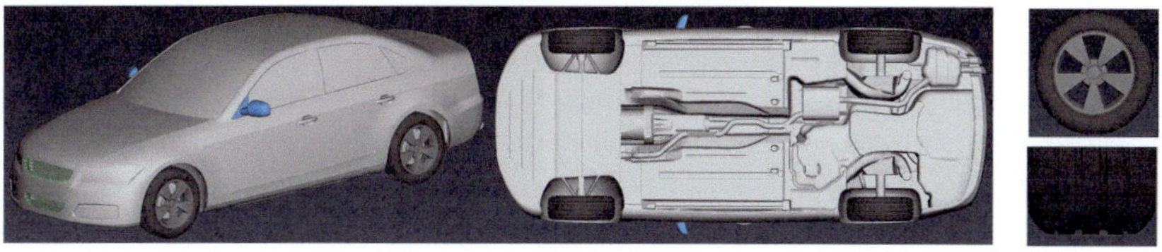

COMMENTS	mirrors included

CFD Results					
Drag (Cx) [-]	0,284	Radiator Mass Flow [kg/s]	1,399	Wheelhouse LHS Flow [kg/s]	0,400
Frontal Area [m^2]	2,170	Upper Grill Flow [kg/s]	0,431	Wheelhouse RHS Flow [kg/s]	0,331
Front Lift (Czf) [-]	-0,029	Lower Grill Flow [kg/s]	0,971	Tunnel Flow [kg/s]	0,659
Rear Lift (Czr) [-]	0,085				
Underhood Ref Pressure (#415) (Cp) [-]:		-0,091	Wheel-house Ref Pressure (#566) (Cp) [-]:		-0,174

Upperbody Centerline Pressure (Cp)

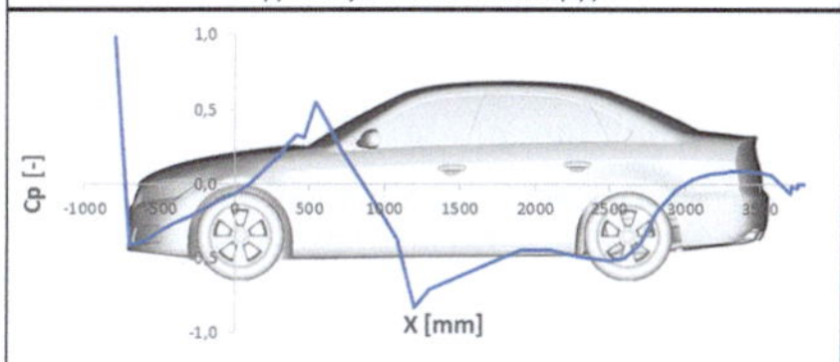

Underbody Centerline Pressure (Cp)

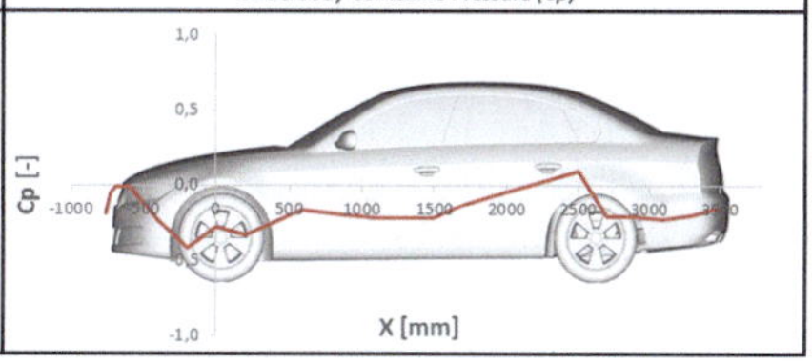

Fig. 25. Open Cooling DrivAer CFD result reporting template

References

1. Ahmed, S., Ramm, G., Faltin, G.: Some salient features of the time-averaged ground vehicle wake. SAE Technical Paper 840300 (1984). https://doi.org/10.4271/840300
2. Closed-Test-Section Wind Tunnel Blockage Corrections for Road Vehicles. SAE Report SP-1176, January 1996. ISBN: 1-56091-815-2
3. Windsor, S.: The effect of rear end shape an road vehicle aerodynamic drag. C427/6/031, IMechE Autotech, UK (1991)
4. Le Good, G., Garry, K.: On the use of reference models in automotive aerodynamics. SAE Technical Paper 2004-01-1308 (2004). https://doi.org/10.4271/2004-01-1308
5. Heft, A., Indinger, T., Adams, N.: Introduction of a new realistic generic car model for aerodynamic investigations. SAE Technical Paper 2012-01-0168 (2012). https://doi.org/10.4271/2012-01-0168
6. Theissen, P., Wojciak, J., Heuler, K., Demuth, R., et al.: Experimental investigation of unsteady vehicle aerodynamics under time-dependent flow conditions - Part 1. SAE Technical Paper 2011-01-0177 (2011). https://doi.org/10.4271/2011-01-0177
7. Wojciak, J., Theissen, P., Heuler, K., Indinger, T., et al.: Experimental investigation of unsteady vehicle aerodynamics under time-dependent flow conditions – Part 2. SAE Technical Paper 2011-01-0164 (2011). https://doi.org/10.4271/2011-01-0164
8. Mack, S., Indinger, T., Adams, N., Unterlechner, P.: The ground simulation upgrade of the large wind tunnel at the Technische Universität München. SAE Technical Paper 2012-01-0299 (2012). https://doi.org/10.4271/2012-01-0299
9. Wojciak, J., Schnepf, B., Indinger, T., Adams, N.: Study on the capability of an open source CFD software for unsteady vehicle aerodynamics. SAE Int. J. Commer. Veh. **5**(1), 196–207 (2012). https://doi.org/10.4271/2012-01-0585
10. Collin, C., Mack, S., Indinger, T., Mueller, J.: A numerical and experimental evaluation of open jet wind tunnel interferences using the DrivAer reference model. SAE Int. J. Passeng. Cars Mech. Syst. **9**(2), 657–679 (2016). https://doi.org/10.4271/2016-01-1597
11. Nayeri, C., Strangfeld, C., Wieser, D., Abbassi, M.R., et al.: Experimental methods, unsteadiness and flow control in vehicle aerodynamics. In: IQPC Aerodynamics Conference, Berlin, 28th Nov 2012
12. Strangfeld, C., Wieser, D., Schmidt, H., Woszidlo, R., et al.: Experimental study of baseline flow characteristics for the realistic car model DrivAer. SAE Technical Paper 2013-01-1251 (2013). https://doi.org/10.4271/2013-01-1251
13. Wieser, D., Lang, H., Nayeri, C., Paschereit, C.: Manipulation of the aerodynamic behavior of the driveAer model with fluidic oscillators. SAE Int. J. Passeng. Cars Mech. Syst. **8**(2), 687–702 (2015). https://doi.org/10.4271/2015-01-1540
14. Stoll, D., Kuthada, T., Wiedemann, J., Schütz, T: Unsteady aerodynamic vehicle properties of the driveAer model in the IVK model scale wind tunnel. In: FKFS Conference on Progress in Vehicle Aerodynamics and Thermal Management, Stuttgart (2015)
15. Stoll, D., Schoenleber, C., Wittmeier, F., Kuthada, T., et al.: Investigation of aerodynamic drag in turbulent flow conditions. SAE Int. J. Passeng. Cars Mech. Syst. **9**(2), 733–742 (2016). https://doi.org/10.4271/2016-01-1605
16. Stoll, D., Kuthada, T., Wiedemann, J.: Experimental and numerical investigation of aerodynamic drag in turbulent flow conditions. In: IMechE International Conference on Vehicle Aerodynamics, Coventry, 21–22 Sept 2016
17. Guilmineau, E.: Numerical simulations of flow around a realistic generic car model. SAE Int. J. Passeng. Cars Mech. Syst. **7**(2), 646–653 (2014). https://doi.org/10.4271/2014-01-0607

18. Jakirlic, S., Kutej, L., Basara, B., Tropea, C.: Computational study of the aerodynamics of a realistic car model by means of RANS and hybrid RANS/LES approaches. SAE Int. J. Passeng. Cars Mech. Syst. **7**(2), 559–574 (2014). https://doi.org/10.4271/2014-01-0594
19. Jakirlic, S., Kutej, L., Hanssmann, D., Basara, B. et al.: Eddy-resolving simulations of the notchback 'DrivAer' model: influence of underbody geometry and wheels rotation on aerodynamic behaviour. SAE Technical Paper 2016-01-1602 (2016). https://doi.org/10.4271/2016-01-1602
20. Jungmann, J., Schütz, T., Tropea, C., Jakirlic, S.: Flow past a DrivAer body in a scaled wind tunnel: computational study by a reference to a complementary experiment. In: IMechE International Conference on Vehicle Aerodynamics, Coventry, 21–22 Sept 2016
21. Soares, R., Garry, K., Holt, J.: Comparison of the far-field aerodynamic wake development for three DrivAer model configurations using a cost-effective RANS simulation. SAE Technical Paper 2017-01-1514 (2017). https://doi.org/10.4271/2017-01-1514
22. Simmonds, N., Pitman, J., Tsoutsanis, P., Jenkins, K. et al.: Complete body aerodynamic study of three vehicles. SAE Technical Paper 2017-01-1529 (2017). https://doi.org/10.4271/2017-01-1529
23. Haag, L., Kiewat, M., Indinger, T., Blacha, T.: Numerical and experimental investigations of rotating wheel aerodynamics on the DrivAer model with engine bay flow. In: Proceedings of the ASME 2017 Fluids Engineering Division Summer Meeting, FEDSM 2017-69305
24. Le Good, G., Annetts, I., Quilter, S., Cross, M., Lewis, R.: The design of systematic add-on configuration changes for the DrivAer body and their aerodynamic characteristics. In: IMechE International Conference on Vehicle Aerodynamics, Coventry, 21–22. Sept 2016
25. Barnard, R.H.: Theoretical and experimental investigation of the aerodynamic drag due to automotive cooling systems. Proc. Inst. Mech. Eng. Part D: J. Autom. Eng. **214**(8), 919–927 (2000)
26. Kuthada, T., Pfannkuchen, E., Wiedemann, J.: Evaluation of cooling air drag on a reference body. In: 5th MIRA International Vehicle Aerodynamics Conference. Warwick, UK (2004)
27. Kuthada, T.: Die optimierung von Pkw-Kühlluftführungs- systemen unter dem Einfluss moderner Bodensimulations- techniken; Dissertation. Universität Stuttgart (2006) ISBN: 3-8169-2664-9
28. Cogotti, A.: A parametric study on the ground effect of a simplified car model. SAE Technical Paper 980031 (1998). https://doi.org/10.4271/980031
29. Baeder, D., Indinger, T., Adams, N., Decker, F.: Comparison of numerical simulations with experiments of bluff bodies including under-hood flow. SAE Technical Paper 2011-01-0171 (2011). https://doi.org/10.4271/2011-01-0171
30. Fares, E.; Jelic, S.; Kuthada, T., Schröck, D.: Lattice boltzmann thermal flow simulation and measurements of a modified SAE model with heated plug FEDSM2006-98467. ASME (2006)
31. Kuthada, T., Wiedemann, J.: Investigations in a cooling air flow system under the influence of road simulation. SAE Technical Paper 2008-01-0796 (2008). https://doi.org/10.4271/2008-01-0796
32. Wittmeier, F., Kuthada, T.: Open grille DrivAer model - first results. SAE Int. J. Passeng. Cars Mech. Syst. **8**(1), 252–260 (2015). https://doi.org/10.4271/2015-01-1553
33. Wittmeier, F.: The recent upgrade of the model scale wind tunnel of university of Stuttgart. SAE Int. J. Passeng. Cars Mech. Syst. **10**(1), 203–213 (2017). https://doi.org/10.4271/2017-01-1527
34. Kuthada, T., Wittmeier, F., Bock, B., Schoenleber, C., et al.: The effects of cooling air on the flow field around a vehicle. SAE Int. J. Passeng. Cars Mech. Syst. **9**(2), 723–732 (2016). https://doi.org/10.4271/2016-01-1603

35. Kuthada, T.: A review of some cooling air flow measurement techniques for model scale, full scale and CFD. SAE Int. J. Passeng. Cars Mech. Syst. **6**(1), 88–96 (2013). https://doi.org/10.4271/2013-01-0598
36. Cogotti, A.: The new moving ground system of the pininfarina wind tunnel. SAE Technical Paper 2007-01-1044 (2007). https://doi.org/10.4271/2007-01-1044

An Experimental Investigation into the Flow Mechanisms Around an SUV in Open and Closed Cooling Air Conditions

John Pitman and Adrian Gaylard(✉)

Jaguar Land Rover, 53/G/3/1, Banbury Road, Gaydon, Warwick CV35 0RR, UK
agaylar1@jaguarlandrover.com

Abstract. A Range Rover SUV with the cooling apertures open and closed had a range of force, surface pressure and flow-field measurements recorded in the FKFS Aero-acoustic wind tunnel. With modern automotive aerodynamics requiring low drag in a range of operating conditions, the primary motivation was to give insight into the interacting flow mechanisms which lead to differing aerodynamic behaviours depending on the vehicle configuration. Cooling drag, the drag difference between the vehicle with open and closed cooling apertures, is used as a metric to demonstrate a set of complex flow interactions occurring on this vehicle between the front and rear wheel wakes and the base wake.

1 Introduction

In recent years automotive manufacturers have pushed toward increasingly lower drag vehicles for the purposes of emissions compliance and customer fuel economy/range.

All areas of the vehicle are open to consideration by the aerodynamicist and as the drag in absolute terms becomes lower, positive or negative interactions arising from particular configuration aspects such as the cooling or wheels become more significant.

Traditionally it might have been possible to assume that development of an exterior body feature for aerodynamics could have been carried out independently of the cooling condition. Indeed this approach has long been taken in order to progress with the exterior body shape in the early stages of a programme when such aspects are often considered too immature to enable the creation of a representative prototype or CFD model.

Experience has shown however, that reliable development of many aspects of modern automotive aerodynamics needs to take into consideration all realistic operating configurations of the vehicle. Active grille shutters are now a common feature on many vehicles, reducing the amount of cooling air flow being allowed through the cooling apertures. In order that the maximum benefit of such features be realised, the rest of the vehicle needs to be developed in open *and* closed cooling conditions.

CFD has become a valuable part of the development process and sits alongside the wind tunnel in the aerodynamicist's toolkit [1]. Conceptual and early stage vehicles may be taken from a scan of a clay model into a representation of a fully engineered vehicle for simulation in CFD. Surrogate or early versions of engine bay and other components can be used to enable a simulation of what is, at the time, the best

J. Wiedemann (ed.), *Progress in Vehicle Aerodynamics and Thermal Management*,
https://doi.org/10.1007/978-3-319-67822-1_4

representation of the vehicle in its finished state. Uncertainty arises however when direct correlation exercises illustrate that predicting the difference in drag between open and closed cooling conditions using CFD can be a very difficult task [2, 3].

An area of significant focus in CFD modelling development in recent years has been the complex issue of a loaded, rotating, treaded tyre in contact with the ground. Immersed boundary methods have recently become available in a number of commercial CFD codes [4]. This enables a fully detailed treaded tyre to be rotated, thus reducing the number of modelling simplifications compared with previous methods. It is fair to say however, that this aspect is still very much a work in progress. The correct prediction of local flow structures around the wheels and tyres is an important pre-requisite in correctly predicting the flow-field around the rest of the vehicle.

The difference between open and closed cooling is generally referred to as 'cooling drag'. Numerous studies, see [5] for example, have shown that this term as defined below, does not purely describe the momentum loss through the engine bay.

$$\text{Cooling drag} = C_D\,\text{Cooling Open} - C_D\,\text{Cooling Closed} \tag{1}$$

CFD or wind tunnel flow visualisation enables the aerodynamicist to see that the flow around a vehicle changes dramatically when the apertures are opened or closed. Separation points and wake regions can change significantly, meaning the exterior flow changes have the potential to add to or offset the engine bay momentum losses.

In recent years Jaguar Land Rover has begun to conduct its aerodynamic development at the FKFS Aero-Acoustic Wind Tunnel (AAWT) in Stuttgart, Germany. The present study looks at the flow field around a Range Rover, in a number of different configurations which exhibit interactions in the FKFS AAWT. The motivations are thus summarised as follows:

- To understand the flow mechanisms that cause significant cooling air flow interactions between vehicle configurations such that they may be exploited for future drag reduction
- To generate a set of flow-field data for comparison with CFD simulation and on-going toolset development
- To isolate the influence of the wheels and their contact with the ground by removing them completely

2 Test Set-up

2.1 Test Vehicle

The test vehicle for this study was a 13MY Supercharged Range Rover. This specification vehicle is generally fitted with wheels ranging from 20 to 22" and does not have active grille shutters. The vehicle is installed into the FKFS AAWT by means of struts which clamp onto the sills as shown in the figure below. The wheels are in loaded contact with the wheel drive units. The aerodynamic forces and moments are thus jointly transferred to the balance via the struts and wheel drive units (Fig. 1).

Fig. 1. Range Rover in baseline configuration - open cooling with 22” wheels

2.2 Test Configurations

The baseline configuration of the vehicle is shown in the figure above – open cooling with a 22” wheel style. A test matrix was formed from combinations of the following variables:

- Open and closed cooling – achieved by taping over all the front apertures: (Fig. 2)

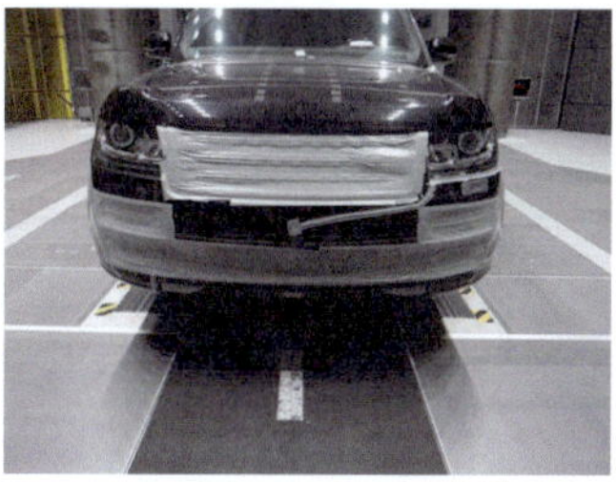

Fig. 2. Closed cooling configuration

- 22” wheels as shown in 1, an alternative 20” style (Fig. 3) and the wheels removed completely (Fig. 4), with the vehicle supported only by the struts. Foam card covers were fitted to each wheel arch to avoid unsteady shedding that may have occurred from the empty wheel-houses.

Fig. 3. 20” wheel style

Fig. 4. No-wheels configuration

- Wheels rotating and static. To isolate the effect of the wheel rotation as much as possible, with the wheels static, the centre belt and all other aspects of the moving ground simulation system were kept as per the rotating configuration
- Ride height: nominal and +50 mm
- Wind speeds of 80 and 140 km/hr in addition to the default 100 km/hr.

2.3 FKFS AAWT

The FKFS AAWT has been extensively developed for automotive aerodynamic purposes over several decades [6]. For the present testing, the wind tunnel was used in 5 belt configuration with all the default ground simulation settings. Test speed is typically 140 km/hr however for this testing the speed was set to 100 km/hr in order to match that used in CFD for computational effort reasons. The difference in Reynolds number is not the main concern, rather the change in shape and position of the rotating tyres due to centrifugal forces. Testing at this lower speed has the additional benefit of reducing the potential for movement of the traverse mounted flow measurement probes (Fig. 5).

Fig. 5. Surface pressure measurements on the tailgate and rear screen

2.4 Measurement Equipment

For this test, 65 surface pressure spades were placed around the vehicle in areas of interest. These are summarised into the following areas:

- Stagnation region
- Inside the cooling ducts
- Around the engine
- Along the centreline on the undertrays
- In front of the front and rear wheel deflectors
- Inside the wheel-arches (front, top, rear)
- Behind each wheel on the bodyside (in the area covered by wheel wake)
- On the base:

For flow measurements away from the surface, a pair of traverse-mounted 12-hole omni-probes was used. These were mounted at 45° in the X-Y plane to enable measurement underneath the car (particularly behind the wheels) from the left hand side as shown in Fig. 6. The traverse moves at a rate of 50 mm per second, averaging the values recorded over each second of data. The omni-probes do not record flow from within a semi-cone angle of 60° around the support axis and hence flow from the negative X-direction and Y-directions is not seen. This characteristic means care must be taken in interpretation of the data recorded, particularly when attempting to compare directly with CFD.

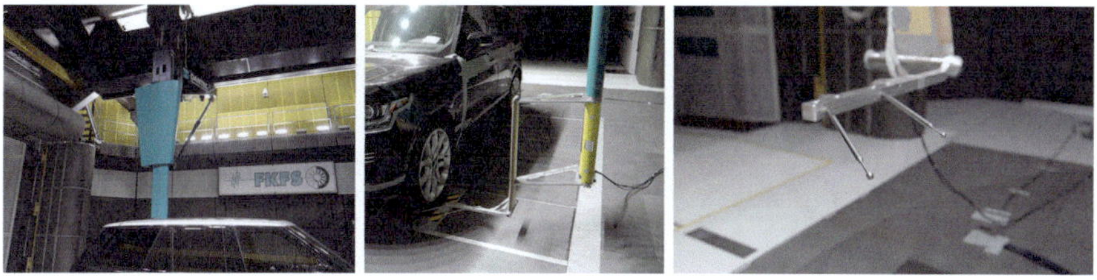

Fig. 6. Traverse mounted omni-probes

The other important consideration is the intrusiveness of the traverse and probe mounting frame. Whilst streamlined, the traverse does have an influence on the flow-field around the vehicle, particularly when in certain positions. Whilst it is not the focus of this study, this interference will be briefly documented in the results section.

2.5 Flow Measurement Locations

CFD was used ahead of the test to determine suitable sizes and locations for the measurement regions around the Range Rover (Fig. 7):

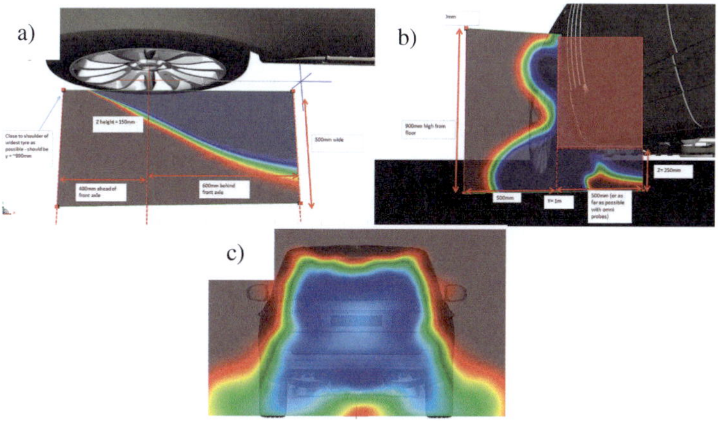

Fig. 7. Wake survey location details: (**a**) FW1: X-Y plane adjacent to the front wheel, 150 mm above the floor; (**b**) FW2a,b: Y-Z planes adjacent to and behind the front wheel respectively, 370 mm behind front axle; (**c**) BW: Y-Z plane 500 mm from the vehicle base

2.6 Measurement Methodology

As a degree of intrusiveness was anticipated from the traverse equipment [7] and pressure loom, balance forces and surface pressures were measured before the start of each flow-field measurement to provide, as best as possible, the drag and surface pressures corresponding to the flow-field being measured. A number of configurations were also measured with the traverse moved away from the car, as shown in Fig. 8:

Fig. 8. Traverse away from car in 'park position'

Prior to analysis, the forces and pressure data were all corrected using the Mercker-Cooper two-gradient method, as described in [8].

3 Results and Analysis

3.1 Initial Observations

Table 1 lists the configurations tested in open and closed cooling conditions, ordered by increasing cooling delta:

The first observation was that the configurations with static wheels exhibited the lowest 'cooling drag'.

Figure 9 shows that, relative to the static wheels configurations with lowest cooling drag, the higher cooling drag results are obtained through varying combinations of increases/decreases in the absolute values of open and closed drag.

Table 1. Test configurations and cooling delta results

Wheels	Ride height	Speed	Rotation	Traverse pos	Δ CD_Cooling
22"	845mm	100 km/hr	No	FW1	0.011
22"	845mm	100 km/hr	No	BW2	0.014
20"	845mm	100 km/hr	No	FW1	0.018
20"	845mm	100 km/hr	Yes	FW2b	0.018
20"	845mm	100 km/hr	Yes	FW1	0.019
20"	845mm	100 km/hr	Yes	FW2a	0.019
20"	845mm	100 km/hr	Yes	RW	0.020
None	895mm	100 km/hr	n/a	Park position	0.020
22"	845mm	100 km/hr	Yes	RW	0.020
20"	845mm	140 km/hr	Yes	FW1	0.021
22"	845mm	100 km/hr	Yes	FW2b	0.022
22"	845mm	80 km/hr	Yes	FW1	0.022
22"	845mm	100 km/hr	Yes	FW1	0.022
20"	845mm	80 km/hr	Yes	FW1	0.022
22"	895mm	100 km/hr	Yes	Park position	0.023
22"	845mm	140 km/hr	Yes	FW1	0.024
22"	845mm	100 km/hr	Yes	Park position	0.024
22"	845mm	100 km/hr	Yes	FW2a	0.024
22"	845mm	100 km/hr	Yes	BW2	0.025
22"	845mm	100 km/hr	Yes	BW1	0.025
20"	845mm	100 km/hr	Yes	BW1	0.026

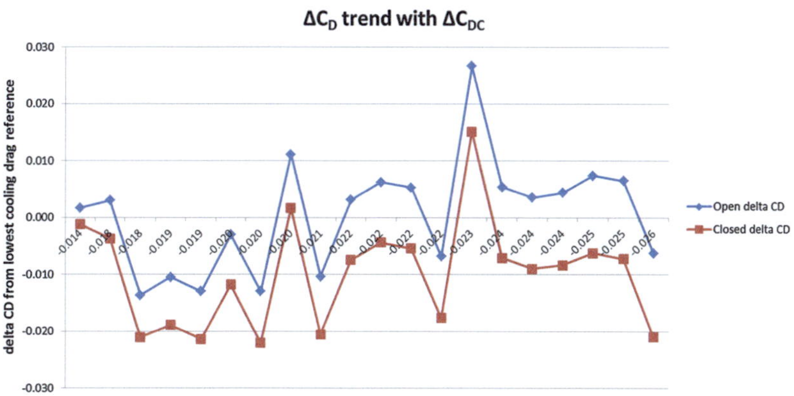

Fig. 9. Trend in delta C_D from lowest ΔC_{DC} reference

3.2 Traverse Influence

The forces were influenced to a varying degree by the location of the traverse. To help visualise this, the base pressure measurements recorded for the open cooling, 22" rotating wheels configuration were plotted using Matlab over an image of the vehicle (Fig. 10):

This effect has been reported by other investigators [7] and is an unavoidable side effect of such flow-field measurement techniques. The pragmatic conclusion was drawn that forces and pressure measurements could only be compared with the traverse in the

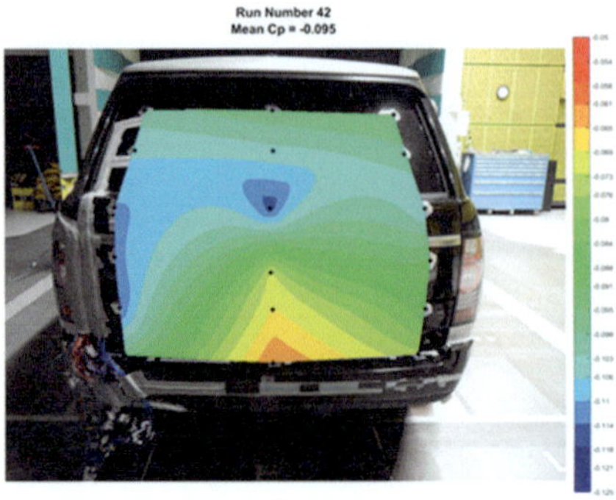

Fig. 10. Base pressure distribution for the reference configuration with wind tunnel traverse in the park position: 22” rotating wheels, open Cooling

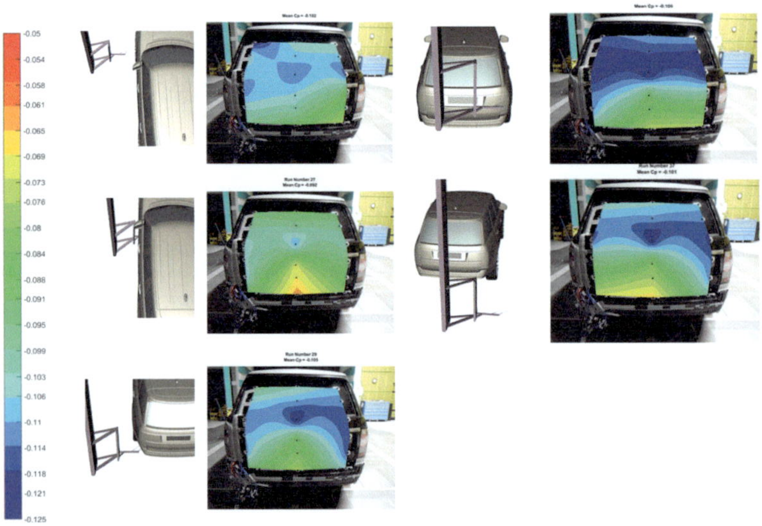

Fig. 11. Influence of traverse position on the base pressure distribution: 22” rotating wheels, open cooling

same location each time. The traverse and probe equipment was scanned and reverse engineered for future CFD correlation work (Fig. 11).

3.3 Effect of Ride Height Change and Removal of Wheels

There was no significant change in cooling drag with the wheels removed completely or the ride-height increased by 50 mm, despite the absolute drag values changing significantly.

Flow-field measurements were not recorded for these configurations, however it was of interest to try and determine, using the surface pressure data, whether the similar ‘cooling drag’ figures were arrived at by the same mechanisms as the more conventional configurations.

As is common practice in automotive aerodynamics, the reference area was kept constant during the test, so C_D essentially becomes analogous to C_DA, which should be used when considering gross changes in the frontal area such as removing the wheels. Table 2 shows the drag increments for increasing the ride height and removing the wheels, alongside the proportion of the total drag on the base and the upper/lower base drag split. This was calculated by integrating the base pressures over their estimated area of influence (as illustrated in Fig. 12), taking into account the slope of the rear screen.

Table 2. Influence of ride height and wheel removal results and base pressure analysis

Ride height, wheels	Cooling	ΔC_D cooling	ΔC_D ride height	ΔC_D wheels OFF	C_D on base	Delta C_D on base	% of Total C_D on base	Change in base % Open to Closed	% base drag on upper (rear screen)	Change in base upper % Open to Closed
845 mm, 22” rotating	Open	0.022	ref	n/a	0.034	−0.004	9.4%	1.9%	40.2%	-0.4%
	Closed		ref	n/a	0.038		11.3%		39.8%	
895 mm, 22” rotating	Open	0.023	0.023	ref	0.039	−0.002	10.2%%	1.2%	39.1%	0.2%
	Closed		0.022	ref	0.041		11.5°%		39.3%	
895 mm, **No wheels**	Open	0.020	n/a	−0.040	0.045	−0.002	13.0%%	1.5%	40.4%	0.2%
	Closed		n/a	−0.037	0.047		14.5%		40.6%	

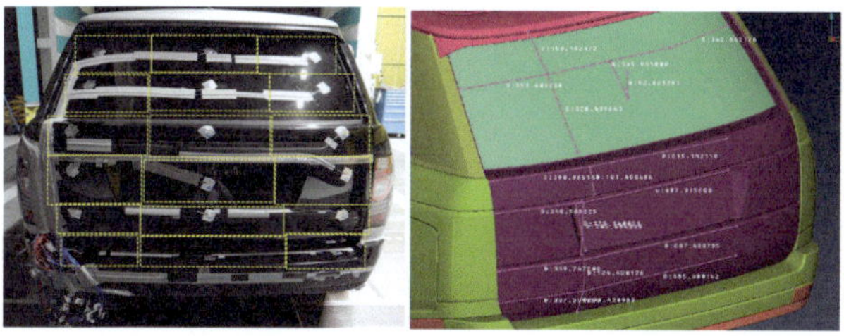

Fig. 12. Base pressure integration areas

This analysis enables the following observations:

- The base drag is higher in each of the closed cooling cases, offsetting to some degree the drag saved at the front of the car by not passing through the cooling pack and engine bay
- Raising the ride height increases the base drag proportion as well as the absolute drag, a good illustration that exposed wheels are very inefficient from a true C_D perspective – i.e. the total C_D increase does not come just from increased drag on the wheels

- In each pair of open/closed tests the greater base drag in the closed case leads to a 1–2% increase in the proportion of drag on the base in the closed vs. the open. The test with the wheels removed unsurprisingly has the greatest proportion of its drag on the base however the increase from open to closed is similar to the normal vehicle
- The drag decrease from removing the wheels was moderated significantly by the position of the fully drooped suspension components. The Range Rover has air suspension which would have required a significant amount of workshop time to enable the axle heights to be kept in their standard position relative to the ground
- The split of base drag between the upper and lower halves varies only slightly around 40/60 for each configuration.

Given the repeatability of the test (± 0.001 C_D for this vehicle) and the relatively small increments in 'cooling drag' due to other changes, it was clear that the wheel rotation effect was most suitable for further investigation of the exterior flow interactions which were of primary interest.

3.4 Influence of Static/Rotating Wheels on Cooling Drag

The following sections present a selection of the surface pressure and flow-field data which was found to best illustrate the flow mechanisms responsible for the net drag changes measured.

3.4.1 Surface Pressure Analysis

Base Pressure

Figure 13 shows that, as with the other configurations presented earlier, the base pressure is lower in the closed cooling cases. What is most interesting is that there is a larger difference between open and closed cooling in the case where the net drag difference is smaller.

Table 3 shows that the greater difference in base drag in the static wheels case almost entirely explains the lesser net difference in balance-measured cooling drag. It appears that the outlier is the open-cooling static wheels case. From here it is of primary interest to understand the mechanisms that enable this.

Body-Side Pressures

Pressure spades mounted in the locations shown in Fig. 14 behind the front and rear wheels show that there is a similar delta between open and closed cooling at the front wheels for both rotating *and* static wheels, however the absolute values are lower with static wheels. This indicates that the static front wheel wake extends further along the bodyside. At the locations behind the rear wheels, a difference can be seen between open and closed cooling with static wheels whereas the corresponding measurements are very similar for rotating wheels.

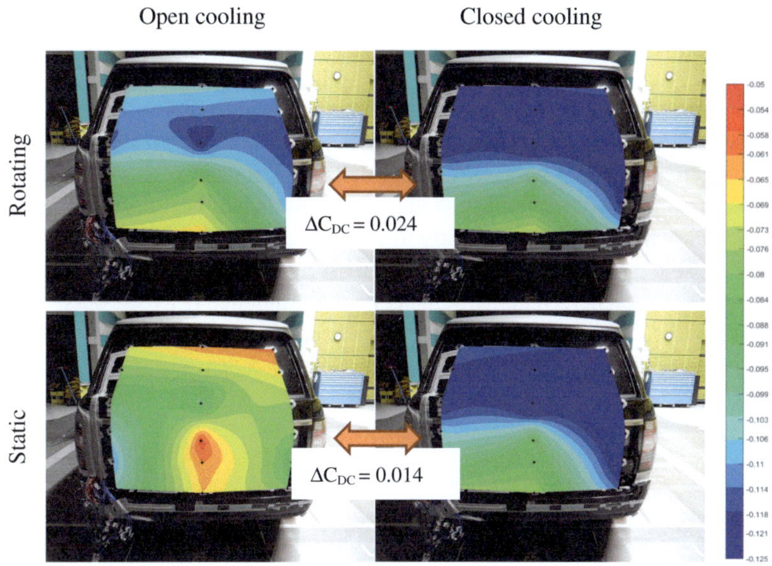

Fig. 13. Base pressure comparison for Open/Closed cooling and Rotating/Static wheels, relating to High/Low cooling drag respectively

Table 3. Rotating/Static wheel base drag integration

Wheels	Cooling	C_D base area	Delta	C_D tailgate	Delta	C_D Rr screen	Delta	% Total C_D on base area	% change open to closed	% Base on Glass (upper)
Rotating	Open	0.036	−0.004	0.021	−0.002	0.015	−0.002	9.9%	1.9%	40.8%
	Closed	0.040		0.024		0.016		11.8%		40.6%
Static	Open	0.027	−0.013	0.018	−0.006	0.009	−0.007	7.6%	4.0%	34.3%
	Closed	0.040		0.024		0.016		11.6%		40.5%

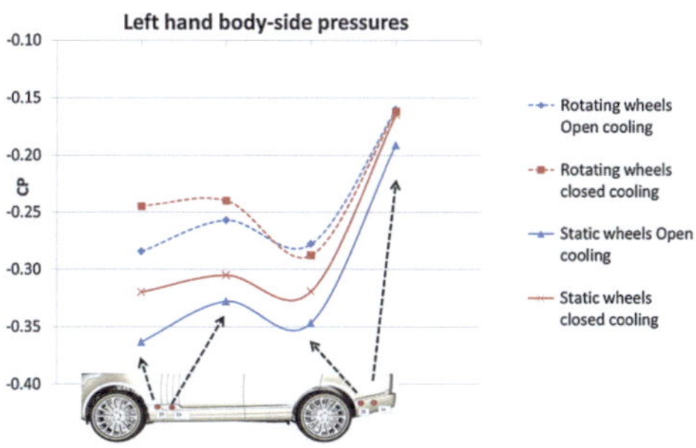

Fig. 14. Body side pressure comparison for Open/Closed cooling and Rotating/Static wheels, relating to High/Low cooling drag respectively

Rear Wheel Deflector Pressures

The rear wheel deflectors sit at the front lower edge of the rear wheel-arches, ahead of the rear wheels. The following pressure measurements are taken at the location shown in the inset figure:

The pressures in front of the rear wheel deflectors appear to be good indicators of differing upstream flow changes which are then responsible for differing flow onto and around the rear wheels and subsequently into the base wake. With rotating wheels, both open and closed cooling have a positive pressure at this location, with the closed cooling ~0.04 C_P higher. With static wheels, the difference between open and closed cooling is double, at ~0.08 C_P and it can be seen that in the open cooling case the pressure is negative (Fig. 15).

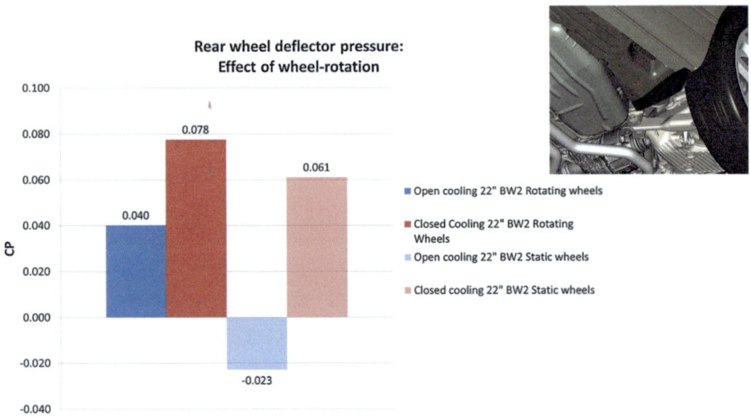

Fig. 15. Rear wheel deflector pressure comparison for Open/Closed cooling and Rotating/Static wheels, relating to High/Low cooling drag respectively

3.4.2 Flow-Field Survey Data

Base Wake Analysis

Figure 16 shows the separated base wake region 500 mm downstream of the vehicle. The overlaid lines delineate a difference in the extent of the wake at the sides for the static cases, between 500 to 1000 mm above the floor, at around Y = ±800 mm. It appears the bulk of the wheel wake is generally lower to the ground in the static cases than for the rotating cases.

Y-velocity was plotted on a vertical line through this region. Figure 16 shows that the static wheels cases have a greater amount of inwash in this area than the rotating wheels cases. The static open cooling case then also has less inwash in the upper half of the wake (Fig. 17).

Figure 18 shows that there is also less downwash in the upper half of the wake for the open-static case than the others. Figure 19, showing the vorticity component of

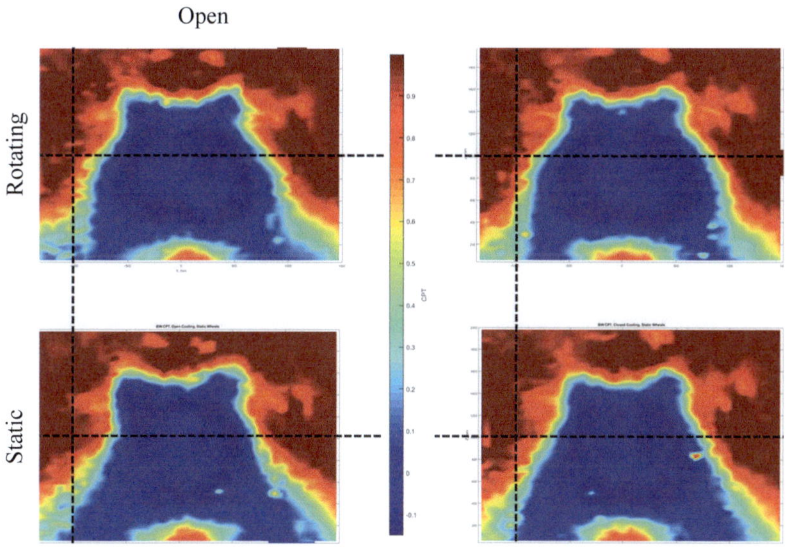

Fig. 16. Base wake (C_{PT}) comparison for Open/Closed cooling and Rotating/Static wheels, relating to High/Low cooling drag respectively

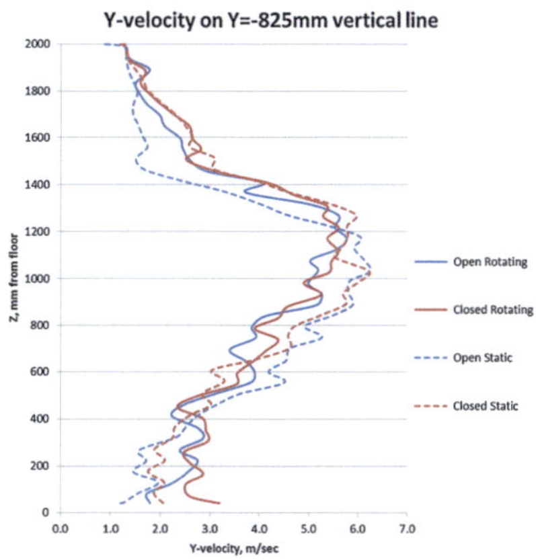

Fig. 17. Comparison of Y-velocity profiles at Y = -825 mm for Open/Closed cooling and Rotating/Static wheels, relating to high and low cooling drag configurations respectively

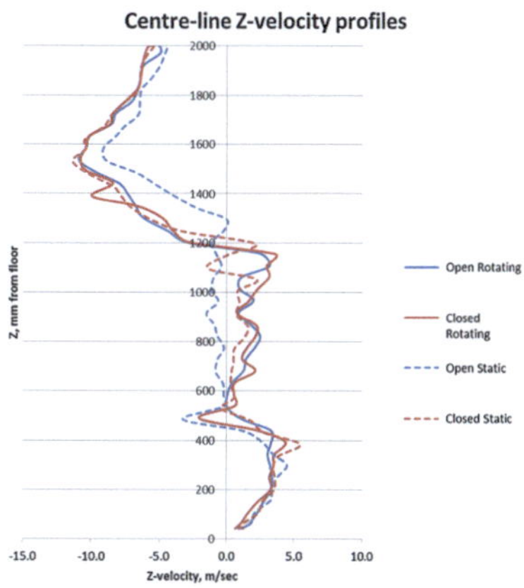

Fig. 18. Centre-line Z-velocity profile comparison

local drag/micro-drag (after Cogotti, [9], the formulation of which is listed in the appendix), is a good illustration of the rotation in the base wake for each configuration, which corresponds well with the base pressure differences seen in Fig. 13. Each closed

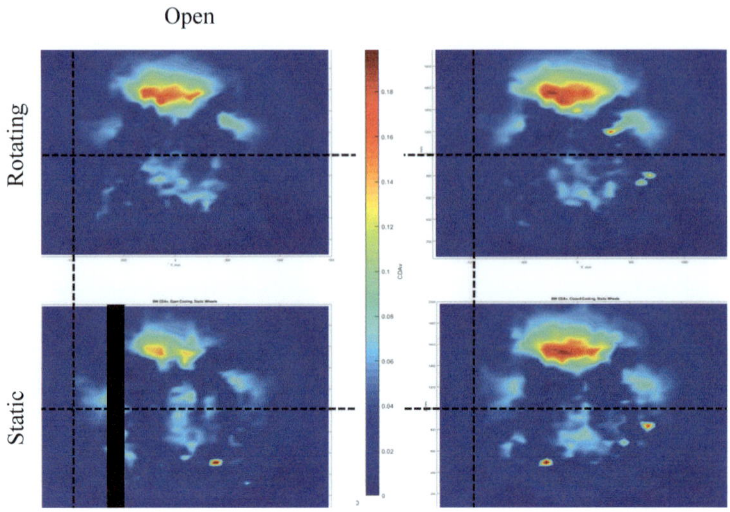

Fig. 19. Base wake vorticity drag comparison for Open/Closed cooling and Rotating/Static wheels, relating to High/Low cooling drag respectively

cooling case has more rotational energy being dissipated in the upper half of the wake and there is a more pronounced difference between the static wheels cases than with rotation.

The centreline velocity profile suggests the reduced downwash in the open-static case is not due to a greater amount of upwash from the underfloor and that the differing mixing behaviour of the rear wheel wake and body-side flow is the main driver.

Front Wheel Wake Analysis

In Fig. 20, the X-Y plane slices adjacent to the front wheel show slightly lower C_{PT} levels within the front wheel wakes for the closed cooling cases than the open. This appears to follow intuition as the closed cooling case will have higher momentum flow hitting the front tyres. The difference is more pronounced due to wheel rotation, with the static wheel wake extending further outward in Y than the rotating wheel wake, as depicted by the edge of the shear layer. This is in line with expectation as the lower half of a rotating tyre surface has a smaller differential in speed to the onset flow than the static tyre.

Figure 21 presents data on FW2, the Y-Z plane behind the front wheel. The property shown is local drag. This property is useful for quantifying the losses in separated wake regions due to momentum and vorticity and can be integrated over a given area to give the drag contribution. This data was only measured for the rotating 22" wheels case but nonetheless adds insight into the differences between open and closed cooling.

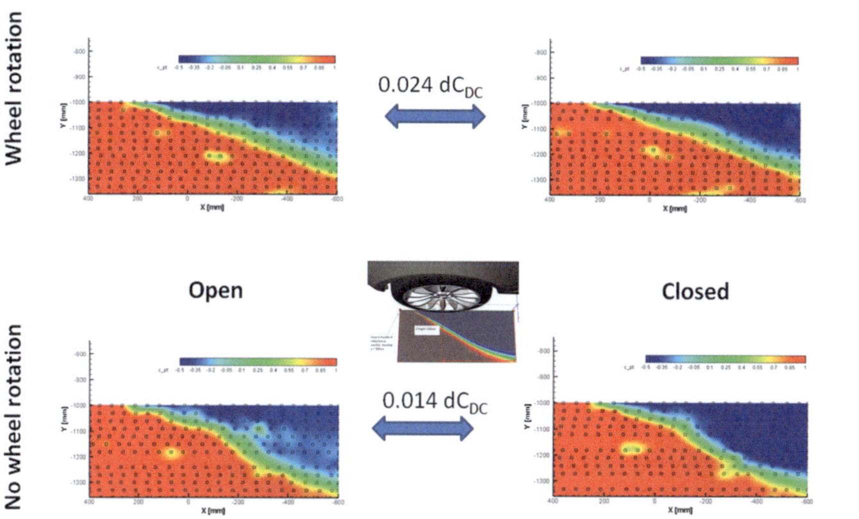

Fig. 20. FW1 C_{P_Total} slices for Open/Closed cooling and Rotating/Static wheels – relating to High/Low cooling drag respectively

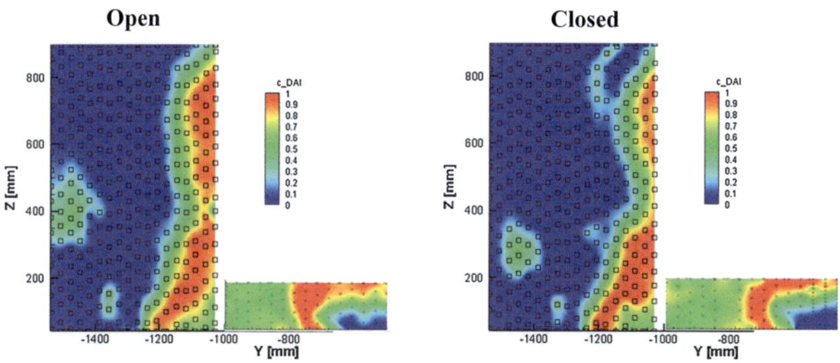

Fig. 21. Local /Micro drag on FW2a,b for open (left) and closed (right) cooling conditions, with 22" wheels rotating

It can be seen that the open cooling case has a larger 'upper lobe' – adjacent to the upper half of the wheel, than the closed case. The closed cooling case has greater losses in the lower half, in-line with observations on FW1.

Integrating across both FW2 slices in Table 4, it can be seen that the larger upper lobe in the open cooling case more than compensates for the reduced losses in the lower half, with double the local C_D contribution on the bodyside slice. Slice (b), underneath the car, also has greater local drag in the open cooling case, with the total for (a) + (b) summing to nearly double that of the closed case.

Figure 22 gives insight into the differing structures within the separated front wheel wakes. Positive Y velocity, towards the centre line, is depicted in red and it can be seen

Table 4. Local drag integrals over FW2 slices for Open and Closed cooling, 22" rotating wheels

FW2 Local drag integration (C_D)	C_D Bodyside slice (a)	C_D Underside slice (b)	FW2 a + b C_D Total
Open cooling	0.033	0.025	0.058
Closed cooling	0.015	0.015	0.031

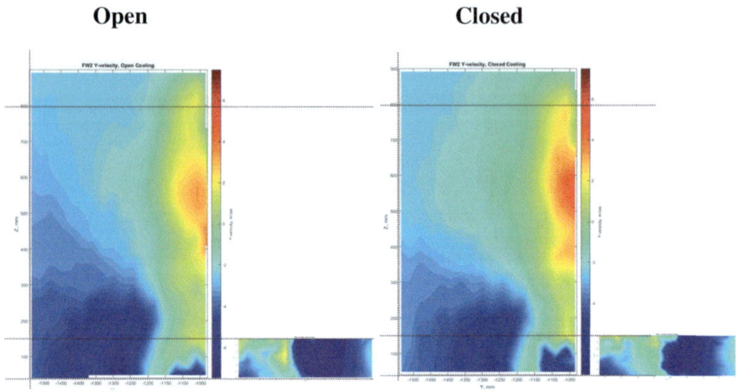

Fig. 22. Y-velocity FW2a,b for open (left) and closed (right) cooling conditions, with 22" wheels rotating. Red = inwards towards Y = 0, Blue = outward

that this inward flow mechanism is stronger for the closed cooling case. Near the ground, there is a stronger outward flow from the inboard edge of the tyre in the open case. These open/closed trends were also seen on the 20" wheel, suggesting the momentum difference due to the flow paths taken is the main driver.

3.4.3 Cooling Drag Mechanism Hypothesis

Analysis of a range of quantitative and qualitative wind tunnel data has enabled a picture to form of the process by which the apparent 'cooling drag'of this vehicle has been altered by an aspect unrelated to the cooling.

On this vehicle, in this wind tunnel, it was seen that the base drag was generally greater for the closed cooling cases. At one level, this follows intuition as when the cooling apertures are closed, the air which would otherwise lose some momentum going through the engine bay all passes around the exterior, thus having a greater potential for creating drag elsewhere.

Stopping the wheel rotation in this case was used as a surrogate for a configuration change that causes a difference in the behaviour of the wheel wakes, such as a front bumper or wheel deflector design change.

Analysis of the size and structure of the front wheel wakes and surface pressures at a key location in front of the rear wheels leads to the hypothesis that the changes at the base were initiated by a chain of events that begins at the front wheels and is likely specific to vehicles with large wheels and SUV ground clearances (Fig. 23).

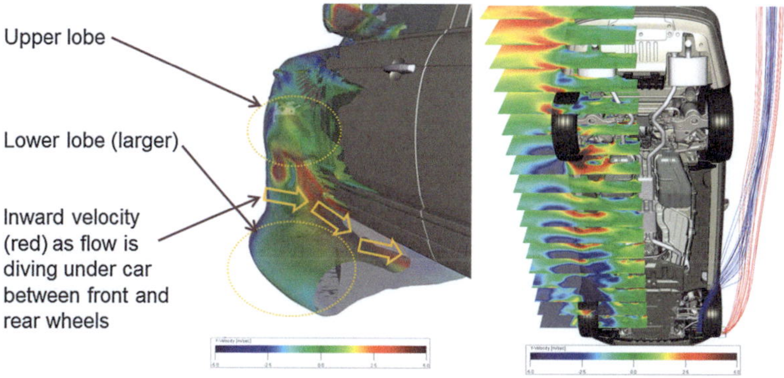

Fig. 23. Range Rover CFD simulation depicting generic flow structures (Y-velocity: blue = outboard; red = inboard)

A generic Range Rover CFD simulation is used to help illustrate the key flow structures:

- A wide front tyre displaces a large amount of fluid, setting up a strong horse-shoe vortex near the ground, which decreases the pressure outside of the tyre further.
- Due to the local yaw angle of the flow onto the front tyre and the horsehoe vortex on the outside, flow from the inboard edge is drawn outwards near the ground. The SUV ride height then enables flow from higher up to roll inwards over the top of this structure and pass under the sill.
- In this analysis, the pressure change ahead of the rear wheel deflectors corresponded with changes in the size of the lower part of the front wheel wake. The size and direction of the front wheel wakes in absolute terms was seen to be the main driver for effects downstream that influenced the difference in drag between open and closed cooling, rather than the difference in the sizes of the front wheel wakes between open and closed cooling.
- The front wheel wake in the static cases spreads out wider, however the area inside the lower wake has reduced losses with the cooling open. This particular wheel wake appears to extend sufficiently far downstream that it shields the rear wheels favourably, indicated by a negative pressure in front of the rear wheels. Conversely, the closed cooling, rotating wheels case appears to draw flow inward and under the sills more strongly, bringing more momentum to the rear wheel area which creates a wheel wake that mixes with the base wake higher up in Z, meaning more of the energy has to be dissipated by the mixing mechanisms in the main body of the wake.
- This front > rear wheel wake mechanism is then responsible for the differing base wake structure changes between open and closed cooling which offset, by a differing amount, the higher drag front wheel wakes and cooling pack momentum losses in the open case. It can be seen that subtle changes to aspects upstream can have amplified effects downstream and the intuitive expectations for what will happen may not hold.

4 Summary and Conclusions

At the outset, the motivation of this study was to understand the flow mechanisms which enable significant interactions in terms of measured drag, such that they may be exploited. The drag difference between open and closed cooling was used as a metric for investigating this, particularly due to the importance of developing vehicles in a range of cooling conditions and the associated challenges with supporting this in CFD.

Whilst likely specific to this vehicle and wind tunnel, the effect of wheel rotation on the difference between open and closed cooling was used to illustrate that changes made around the front wheels can have differing influences downstream depending on how the vehicle is configured.

Aerodynamic development processes at a local level will remain the same – controlling separation, consideration of downstream effects etc., however opportunities for drag reduction and robustness across a range of driving conditions lie in dilligent development strategies involving (particularly) differing wheel and cooling configurations.

On this vehicle, a notable increase in base pressure resulted from subtle changes and interactions which started a long way upstream. CFD is an invaluable tool for visualising such mechanisms and a key learning point from this study, related to the development of its predictive capability, was that the size and behaviour of flow structures in absolute terms is very important. This may mean that resolution increases or modelling changes such as the inclusion of wind tunnel boundary elements, which in isolation have no apparent benefit, are actually necessary pre-requisites for the prediction of complex interactions, such as occur between open and closed cooling on a given vehicle.

Acknowledgements. The author wishes to thank Domenik Schramm, Felix Wittmeier and Timo Kuthada at FKFS for consultation on the flow-field surveys, James Imrie and Rick Shock for supporting the experimental work, and Sebastien Chaligne for proof-reading and general consulation, and Jaguar Land Rover for the permission to publish this work.

Appendix

Micro-drag formulation, after Cogotti [9]:

$$C_D.Aref = \int_A (1 - C_{P_{TOT}})dA - \int_A (1 - \frac{V_X}{V_\infty})^2 dA + \int_A (\frac{V_Y}{V_\infty})^2 dA + \int_A (\frac{V_Z}{V_\infty})^2 dA$$

Where:

Aref = projected frontal area
C_{P_Tot} = coefficient of total pressure
Vx = velocity component in the x-axis
Vy = velocity component in the y-axis

Vz =	velocity component in the z-axis
V =	freestream velocity
A =	wake plane area

References

1. Gaylard, A.: The Appropriate Use of CFD in the Automotive Design Process. SAE Technical Paper 2009-01-1162 (2009). https://doi.org/10.4271/2009-01-1162
2. Lietz, R., Larson, L., Bachant, P., Goldstein, J., et al.: An Extensive Validation of an Open Source Based Solution for Automobile External Aerodynamics. SAE Technical Paper 2017-01-1524 (2017). https://doi.org/10.4271/2017-01-1524
3. Larson, L., Gin, R., Lietz, R.: Aerodynamic investigation of cooling drag of a production sedan part 2: CFD results. SAE Int. J. Passeng. Cars Mech. Syst. **10**(1), 214–223 (2017). https://doi.org/10.4271/2017-01-1528
4. Lew, C., Gopalaswamy, N., Shock, R., Duncan, B. et al.: Aerodynamic Simulation of a Standalone Rotating Treaded Tire. SAE Technical Paper 2017-01-1551 (2017). https://doi.org/10.4271/2017-01-1551
5. Tesch, G., Demuth, R., Adams, N.: A new approach to analyzing cooling and interference drag. SAE Int. J. Passeng. Cars Mech. Syst. **3**(1), 339–351 (2010). https://doi.org/10.4271/2010-01-0286
6. Blumrich, R., Widdecke, N., Wiedemann, J., Michelbach, A. et al.: New FKFS Technology at the Full-Scale Aeroacoustic Wind Tunnel of University of Stuttgart. SAE Int. J. Passeng. Cars Mech. Syst. **8**(1), 294–305 (2015). https://doi.org/10.4271/2015-01-1557
7. Sterken, L., Sebben, S., Walker, T., and Lofdahl, L.: Experimental and Numerical Investigations of the Base Wake on an SUV. SAE Technical Paper 2013-01-0464 (2013). https://doi.org/10.4271/2013-01-0464
8. Mercker, E., Cooper, K.: A Two-Measurement Correction for the Effects of a Pressure Gradient on Automotive, Open-Jet, Wind Tunnel Measurements. SAE Technical Paper 2006-01-0568 (2006). https://doi.org/10.4271/2006-01-0568
9. Cogotti, A.: A Strategy for Optimum Surveys of Passenger-Car Flow Fields. SAE Technical Paper 890374 (1989). https://doi.org/10.4271/890374

Unsteady Flow

Evaluation of Unsteady Flow Phenomena Induced by the Tailgate Gap of a Production Car Using Simulations and Experiments

Georg Eitel-Amor(✉), Sascha Riedl, and Reiner Weidemann

Opel Automobile GmbH, Bahnhofsplatz, 65423 Rüsselsheim am Main, Germany
georg.eitel-amor@opel.com

Abstract. The gaps at the tailgate of passenger cars can play an important role in the generation of wind noise. Depending on the particular geometry, flow separation and vortices may occur that cause pressure fluctuations perceived by the passengers. The underlying aeroacoustic effects can be hard to predict and extensive testing is required to eliminate disturbing wind noise.

In the present work, the acoustic characteristics of an exemplary sedan car are investigated with a focus on the rear end. Different sealing concepts for the tailgate gap are evaluated using wind tunnel measurements and computational fluid dynamics. The study incorporates interior-noise spectra and a deeper analysis of pressure fluctuations around the rear glass panels.

The results imply that the numerical method is suitable to predict the performance of different sealing concepts if small gaps and structures are sufficiently resolved. Furthermore, the simulation provides insight to physical mechanisms and allows the localization of sensitive areas.

1 Introduction

When optimizing the exterior shape of a car with respect to aerodynamic drag and wind noise, growing attention is paid on geometrical details. In simulations and wind tunnel measurements it is common practice to consider underbody and suspension parts in order to obtain the right forces and moments acting on the vehicle. However, realistic upper-body details like seals, gaps, and underlying cavities may also influence the performance and the level of comfort experienced by passengers. As a matter of fact hardware testing is only available at a late stage of development, but recent simulation tools enable an evaluation long before the first prototype is built. This allows critical areas to be identified just in time and helps assessing different levels of content.

In this study we perform transient simulations and wind tunnel tests investigating the acoustic properties of the hatch door gap of a mid-size sedan car. Different sealing concepts are assessed and the noise spectra from simulation and test are compared. The aim is to verify the predictive capabilities of numerical simulations which would ease the development process and avoid expensive repair solutions.

The area around the hatch door is of special interest for both aerodynamics and aeroacoustics. Since the tailgate gap disturbs the attached flow towards the rear end of the car, flow separations might increase drag forces and lead to pressure fluctuations

J. Wiedemann (ed.), *Progress in Vehicle Aerodynamics and Thermal Management*,
https://doi.org/10.1007/978-3-319-67822-1_5

acting on the rear window. Furthermore, the flow may excite acoustic modes inside the cavity leading to aggravating resonance effects. It is well known that the vertical positioning of the hatch door has a large influence on the noise level favoring a flush or sub-flush position with respect to the roof's rear end. However, the size and shape of the cavity around the hatch, interaction with upstream flow structures (e.g. A-pillar vortex, antenna wake) and the influence of radii and corners are hard to estimate without numerical studies or hardware tests.

Therefore, this paper describes how time-dependent simulations can be utilized to model the actual acoustic properties of a realistic tailgate geometry. By comparing simulation data with wind tunnel test results, strengths and weaknesses of the model are analyzed. In the following, we will briefly introduce the numerical method and the wind tunnel test procedure. After a description of the case setup, the results of both simulation and test will be presented and evaluated which leads to a final summary and conclusion.

2 Numerical Method

The numerical simulations presented in this study were performed with the commercial software package Exa PowerFLOW® 5.2a. It utilizes the Lattice-Boltzmann Method (LBM) allowing Very Large-Eddy Simulations (VLES) at low Mach numbers [1]. The numerical approach is inherently time-dependent and highly suitable for the efficient solution of aero-acoustic problems [2]. Based on a surface representation of the vehicle and local resolution settings, a Cartesian volume mesh is built during the automated grid generation process. Therefore, complex geometries can be handled efficiently.

The PowerFLOW® suite provides a virtual acoustic wind tunnel with a blockage ratio of 0.1% embedded in an anechoic outer layer. The time-dependent simulation is performed collecting the pressure information in regions of interest (e.g. probes, surfaces or volumes). On measurement surfaces, e.g. the rear window, the cell size is set to 1 mm which leads to a sampling rate of about 20 kHz for the pressure signal. With a simulated physical measurement time of 0.5 s one obtains a well resolved spectrum in a frequency range between 100 Hz and 4 kHz. Furthermore, virtual probes, measurement planes, and measurement volumes can be defined at which time-dependent information is recorded for further post-processing. The overall physical time simulated is 0.75 s, whereas the first 0.25 s are reserved for the initial transient (when starting from a converged solution for the same kind of vehicle).

To determine a noise spectrum for the interior of a realistic car, the obtained pressure fluctuations on the glass panels are processed in a Statistical Energy Analysis (SEA) which models the transfer of sound through the glass panels and considers the acoustic properties of the cabin [3–5]. Further paths of noise transfer from the outside to the interior are neglected in this study. Nevertheless, underbody noise and noise transfer through rubber seals are effects that could be addressed using the same software suite.

3 Test Facility and Measurement Tools

The acoustic wind tunnel tests were performed at FKFS Stuttgart [6]. The interior noise was measured with four microphones located at the driver's and passenger's head position with the following specifications:

- microphones; Brüel & Kjaer ½ in Type 4943
- microphone preamplifiers; Brüel & Kjaer type 2669
- microphone calibrator; Brüel & Kjaer type 4231
- data acquisition hardware system; Headlab labM6 No. 3724

The frequency resolution is 32 Hz and the frequency range is 100 Hz to 10 kHz. Linear averaging is applied with 50% overlap and 500 averages in a measurement period of 10-20 s.

4 Test Case Setup

For the following investigation an Opel Insignia hatchback of model year 2017 was used. In order to investigate the influence of different tailgate sealing concepts at 140 km/h, four variants were defined as shown in Fig. 1:

(a) The car is completely taped leaving no gap around the tailgate (optimum)
(b) The hatch-door cut is completely opened with no seal installed (worst case)

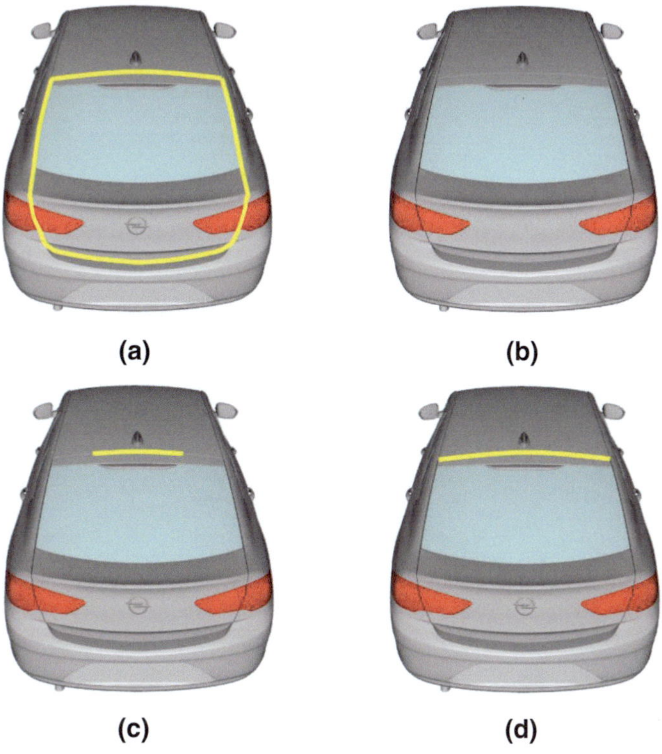

Fig. 1. Description of the investigated variants, where the yellow line indicates where the tailgate gap has been closed: (a) completely closed tailgate gap; (b) completely open tailgate gap; (c) sealing between the hatch-door hinges; (d) sealing of the full upper hatch door cut.

(c) The upper hatch door cut is sealed between the hinges
(d) The upper hatch door cut is sealed completely

In the wind tunnel the different states were realized by adding and or removing adhesive tapes and repeating the standardized data acquisition.

For the numerical study the full body of the car was modelled with a focus on the rear end. All gaps, except for the hatch door cut, were closed representing the taped vehicle as measured at FKFS. The gap around the hatch door and the underlying cavity were carefully transferred to a surface mesh in order represent the whole system as accurately as possible. The smallest voxels of the final Cartesian mesh have a size of 0.5 mm, ensuring that mass flow through thin gaps is sufficiently resolved (see Fig. 2). With the described resolution settings the overall number of voxels was 322 million for the cases with open cavity. Each of the four simulations ran for about 730.000 time steps using 14.000 CPU hours on 288 cores.

To keep the amount of data within reasonable bounds, the simulation only considers noise transfer through the hatch-door glass. This is justified by the assumption that the side-window contribution does not change for the considered variants, i.e. that the simulation will only lack a constant offset in noise level when compared to the measurement.

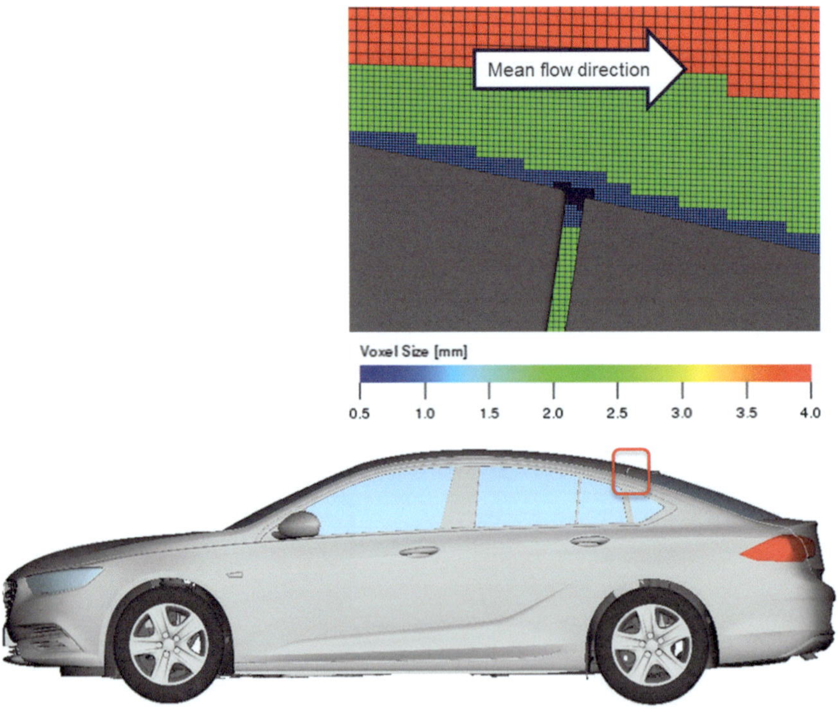

Fig. 2. Side view of the simulation model (bottom) and magnified image of the upper hatch door cut (top). The enlarged picture shows a vertical cut through the Cartesian grid as used in the CFD study. Blue color indicates small voxels and red color larger voxels.

5 Results and Evaluation

The noise spectra obtained from measurements and simulations are shown in Figs. 3 and 4, respectively. For both methods the overall shape of the spectrum is similar showing a decay with increasing frequency and a slightly lower noise level for the simulation. Note that the simulation is expected to be deliver lower noise levels since only the rear windows transfer noise to the interior. Remarkable differences between

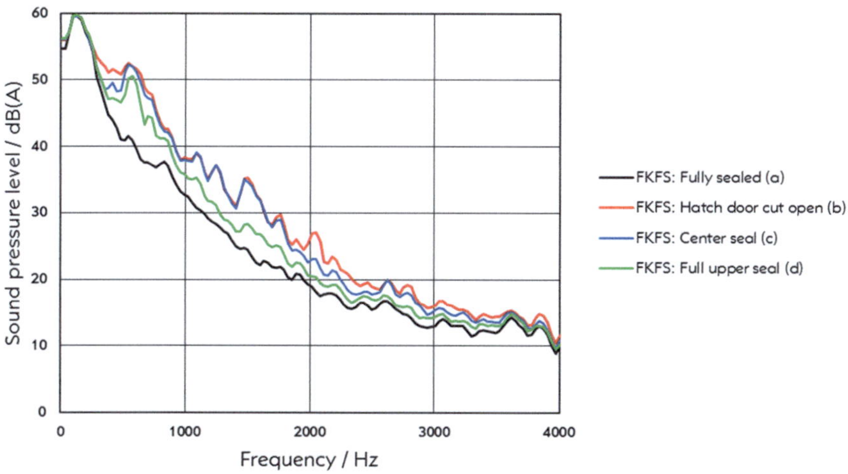

Fig. 3. Noise spectra measured in the FKFS acoustic wind tunnel with a microphone located at the left rear seat.

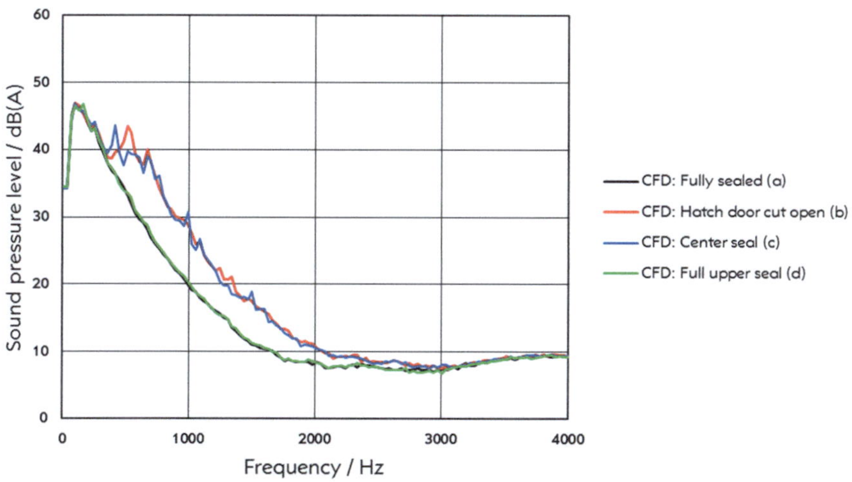

Fig. 4. Noise spectra computed from time-dependent simulations for the left rear seat.

the sealing variants (a)-(d) are observed above 300 Hz and below 4 kHz and will be discussed in the following.

Simulation and experiment evidence a remarkable noise reduction when going from an open tailgate gap to a more and more closed sealing system. As expected, the lowest noise level is observed for the completely sealed tailgate in both methods. The center seal with a width of 600 mm between the hinges brings almost no improvement when compared to the fully open state. This very small potential is confirmed by the wind tunnel test and by the simulation data.

In a next step, the sealing is extended to the full width of the upper hatch-door cut. In the experiment, this leads to a remarkable noise reduction across the whole frequency range shown in Fig. 3, whereas the noise level is still clearly higher than in the optimum case. In the simulation, however, the extended sealing performs as good as the fully sealed variant. That means some physical effect causing interior noise in the

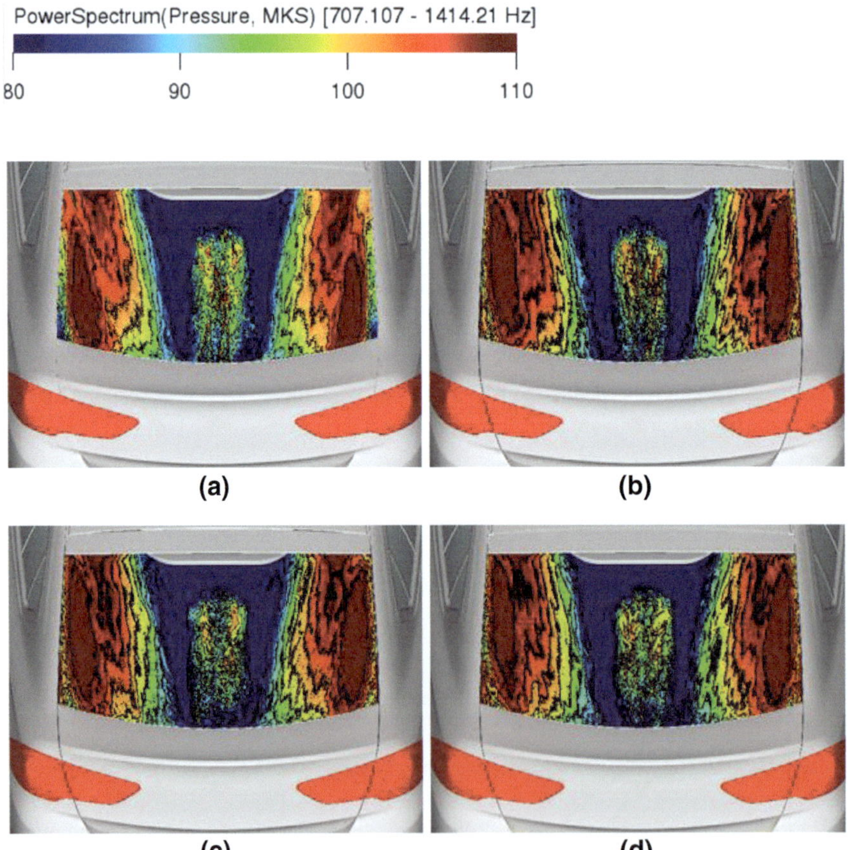

Fig. 5. Pressure dB-maps of the rear glass for different sealing conditions: (a) tailgate gap is completely sealed; (b) tailgate gap is completely open; (c) upper hatch-door cut is sealed between the hinges; (d) upper hatch-door cut is sealed completely. The color code is blue for low pressure amplitudes and red for large pressure fluctuations.

experiment is not captured in the simulation. The reason could be a minor deviation of geometry, locally insufficient resolution, or simply the absence of an additional noise transfer path in the simulation. Since the noise level also depends strongly on the yaw angle, an imperfect alignment in the wind tunnel could also lead to deviations. However, although the sealing performance of variant (d) is overestimated by the simulation, it shows the right trends and both wind tunnel and simulation imply that the outer part of the sealing is the most important one.

Apart from the noise spectra, the computational results allow for the visualization of pressure dB maps on surfaces of interest. Thereby, a deeper understanding of the physics involved can be achieved. In Fig. 5 the distribution of the simulated sound-pressure level on the rear glass is shown for the octave band centered at 1 kHz. These spectra take into account all pressure fluctuations, i.e. hydrodynamic and acoustic ones. All four variants show a 'silent' area in the wake of the stop light and have the highest intensity close the C-pillars. The completely sealed tailgate (upper left picture) stands out because it shows less pressure fluctuations in the upper corners of the rear glass than all other variants.

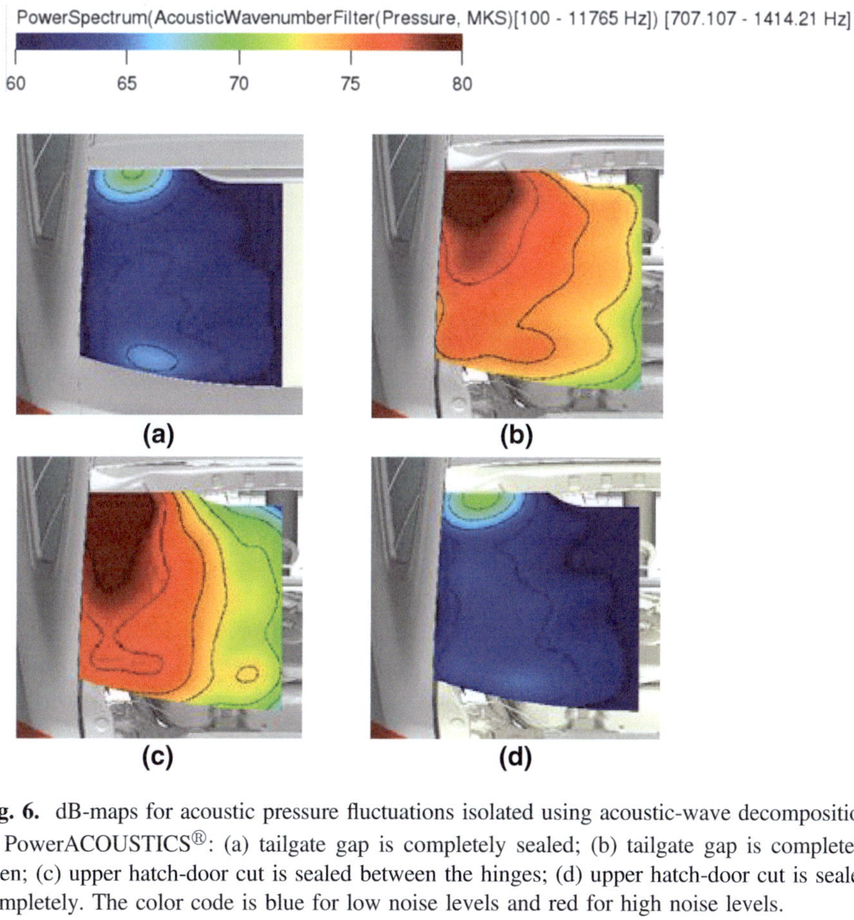

Fig. 6. dB-maps for acoustic pressure fluctuations isolated using acoustic-wave decomposition in PowerACOUSTICS®: (a) tailgate gap is completely sealed; (b) tailgate gap is completely open; (c) upper hatch-door cut is sealed between the hinges; (d) upper hatch-door cut is sealed completely. The color code is blue for low noise levels and red for high noise levels.

However, the good performance of variant *(d)* cannot be explained from this perspective. To understand what causes the improvement in interior noise, one needs to filter out hydrodynamic pressure fluctuations travelling with the convection velocity of the fluid. What remains are acoustic waves travelling at the speed of sound. This filtering step can performed in PowerACOUSTICS® using acoustic-wave decomposition. The dB maps for this filtered component are shown in Fig. 6 clearly indicating a much lower noise level on the whole rear window for variants (a) and (d).

When combining the information of the interior spectra with the aforementioned dB maps, it becomes obvious that the most sensitive area for wind noise optimization of the tailgate is at the outer corners of the tailgate gap (downstream of the roof rails). A closer look at the flow field for variant (b) reveals a flow separation at the outer leading edge of the hatch door and a upward airflow coming from the lower tailgate (see Fig. 7). Closer to the center, the flow passes the tailgate gap smoothly, which explains why a sealing shows almost no effect at this location.

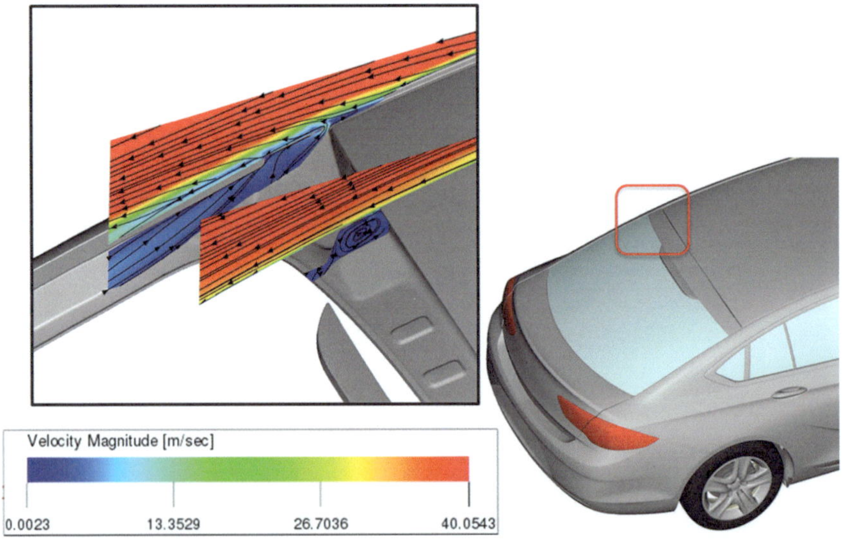

Fig. 7. Velocity field in two vertical cut planes for variant (b) averaged over 0.1 s. The hatch door has been hidden in the upper left picture for reasons of visibility. The flow separation at the outermost corner of the hatch door is clearly visible and an upward suction effect through the side cavity of the tailgate is observed. Closer to the centerline the flow passes the gap without remarkable disturbance. The color code is blue for low velocity magnitudes and red for high velocities.

6 Summary

In the present study we investigated the flow around the tailgate of a production car. The focus was on wind noise being generated by the hatch door gap and transferred through the rear window. To assess the performance of sealing at the upper hatch door cut, we first defined two extreme cases, namely a completely sealed tailgate without any gaps and, on the other hand, a variant with a realistic gap around the hatch door. These cases were compared to a partially and fully sealed upper gap. The interior noise spectra obtained from wind tunnel measurements and from numerical simulations show the same trends, leading to the conclusion that a partial sealing brings almost no improvement for interior noise. Deviations between the absolute numerical and experimental values are a consequence of the simulation model, which focuses on the most relevant flow physics. Taking a closer look into the numerical results, it is found that acoustic waves are mainly generated close to the upper C-pillar and that sealing needs to be applied in this area.

7 Conclusion

Unsteady flow phenomena at the transition from the roof to the tailgate can be critical for interior noise and may require specialized sealing concepts to achieve targets. It has been shown that numerical simulations of the three-dimensional, time-dependent flow field are eligible to predict wind noise phenomena also for intricate geometries with small gaps.

Since the trends and characteristics are correctly predicted, program decisions and geometry changes may be assessed before test hardware is available. Furthermore, the extensive three-dimensional information that comes with the simulations allows for detailed diagnostics and explanation of certain acoustic phenomena observed in hardware tests and helps in finding tailored solutions. As a next step, we plan to improve the agreement of absolute noise levels and answer open questions with regard to wind tunnel alignment and comparable surface measurements.

The presented results underline the potential of time-dependent, numerical simulations for optimization of aero-acoustics in industrial product development. Therefore, we conclude that the presented computational solution complements the conventional experimental approach and will play an important role in Opel's wind noise design process in the near future.

Acknowledgments. We gratefully acknowledge the support of Exa GmbH and the fruitful collaboration with the FKFS Forschungsinstitut für Kraftfahrwesen und Fahrzeugmotoren Stuttgart.

References

1. Kotapati, R., Keating, A., Kandasamy, S., Duncan, B., et al.: The Lattice-Boltzmann-VLES method for automotive fluid dynamics simulation, a review. SAE Technical Paper 2009-26-0057 (2009)
2. Brès, G., Pérot, F., Freed, D.: Properties of the Lattice Boltzmann method for acoustics. In: 15th AIAA/CEAS Aeroacoustics Conference (30th AIAA Aeroacoustics Conference) Aeroacoustics Conferences: American Institute of Aeronautics and Astronautics (2009)
3. Moron, P., Powell, R., Freed, D., Perot, F., et al.: A CFD/SEA approach for prediction of vehicle interior noise due to wind noise. SAE Technical Paper 2009-01-2203 (2009)

4. Lepley, D., Graf, A., Powell, R., Senthooran, S.: A computational approach to evaluate the vehicle interior noise from greenhouse wind noise sources. SAE Technical Paper 2010-01-0285 (2010)
5. Graf, A., Lepley, D., Senthooran, S.: A computational approach to evaluate the vehicle interior noise from greenhouse wind noise sources - Part II. SAE Technical Paper 2011-01-1620 (2011)
6. Blumrich, R., Widdecke, N., Wiedemann, J., Michelbach, A., et al.: New FKFS Technology at the Full-Scale Aeroacoustic Wind Tunnel of University of Stuttgart. SAE Int. J. Passeng. Cars Mech. Syst. **8**(1), 294–305 (2015)

Characterisation of Wake Bi-stability for a Square-Back Geometry with Rotating Wheels

Giancarlo Pavia(✉) and Martin Passmore

Loughborough University, Stewart Miller Building,
LE11 3TU Loughborough, UK
G.Pavia@lboro.ac.uk

Abstract. In this paper the effects produced by the wheels on the bi-stable reflectional symmetry breaking (RSB) mode seen for the wake of a square-back geometry (Grandemange et al. [11]) are investigated considering a modified version of the Windsor body already studied in Perry et al. [18]. The contribution of the wheels and their rotation to the changes in the base pressure distribution and the wake topology is characterised by means of pressure tappings and 2D-3C particle image velocimetry. Balance measurements are used to further characterise the changes in the strength of the RSB mode. For the pure square-back configuration, the results show a general increase of the base drag as a consequence of the strengthening of the suction over the lower portion of base, due to the formation of a pair of counter rotating vortices acting close to the bottom trailing edge. At the same time, the RSB mode is weakened, leading to a reduction in the fluctuations recorded for the lateral component of the aerodynamic force. The sensitivity of the RSB mode to small changes in the shape of the model's trailing edges is characterised by looking at the effects produced by short tapers, with a slant angle of 12° and a chord equal to 4% of the model length, applied to either the horizontal or the vertical trailing edges. The results show that the RSB mode disappears when the effect of the wheels is paired to the upwash generated by the slanted surface (when applied to the bottom trailing edge), although it is still clearly visible when the tapers are applied to the side edges of the base, in contrast with the results reported by Pavia et al. [16] for the same geometry without wheels.

1 Introduction

Square-back vehicles are popular in the automotive industry thanks to their large cabin space and easy access. The large base area, however, yields a higher level of aerodynamic drag compared to other more streamlined rear-end shapes such as those used in fast-back or notch-back cars.

The steady state topology of the wake generated by such geometries has been fully characterised, from both a numerical (Krajnovic and Davidson [14]) and experimental (Perry et al. [17]) point of view. It consists of a toroidal structure which is source of significant pressure losses and whose interaction with the surrounding shear layers results in the formation of a pair of counter rotating vortices downstream of the near

J. Wiedemann (ed.), *Progress in Vehicle Aerodynamics and Thermal Management*,
https://doi.org/10.1007/978-3-319-67822-1_6

wake region. Obtaining accurate results from the numerical simulations of this vortical structure, however, is still an open challenge (Makihara et al. [15]), due to the complex dynamics that accompany the development of the wake itself.

Grandemange et al. [11], testing a 1/4 scale Ahmed body (see Ahmed et al. [1]), pointed out that the wake of a square-back geometry is characterised by the coexistence of two different types of coherent motions. The first is a set of global fluctuating modes, consisting of a lateral and vertical flapping as well as a longitudinal pumping (all with a characteristic time of $T_s \sim 5\,H/V_\infty$ as confirmed by Volpe et al. [23]). The second type of motion is a bi-stable reflectional symmetry breaking (RSB) mode, whose characteristic time is about three orders of magnitude longer than the first one ($T_l \sim 10^3\,H/V_\infty$, with V_∞ the free-stream velocity). The succession of RSB states has been shown to behave like a weak chaotic process with two attractors (Varon et al. [22]). It leads to large variations in the values of side force that have been shown to be responsible for part of the drag experienced by the model (Grandemange et al. [12]).

A similar trend has been also noticed by Grandemange et al. [9] testing an Ahmed body scaled to a full size small car (resulting in a model height based Reynolds number of $Re_H = 2.5 \cdot 10^6$) and equipped with four rotating wheels. When the model was tested for values of the yaw angle Ψ between $-0.5°$ and $+0.5°$, large fluctuations were noticed in the signal related to lateral component of the aerodynamic force recorded at 10 Hz for 40 s, whilst the drag remained constant in the same range.

The same dependency between C_D and Ψ was found by Bonnavion et al. [4] while assessing the presence of bi-stability on the wake generated by a commercially available light van. They reported the existence of a bi-stable regime between two different positions of the reverse flow impinging on the base when the model was at 4° yaw. In this case, however, the RSB mode was found to have an affect mainly on the vertical component of the aerodynamic force, as the model high was greater than the width, in agreement with the conclusions drown by Grandemange et al. [10] studying the effects of different values of the base aspect ratio on the RSB mode.

A deeper understanding of the bi-stable behaviour has been achieved while also developing suitable strategies to control this type of instability. Grandemange et al. [12] suggested that a base drag reduction between 4% and 9% can be achieved for an Ahmed body when the lateral symmetry of the wake is restored by means of a vertical control cylinder positioned in the core of the wake's reverse flow region. An even bigger improvement was reported by Evrard et al. [7] applying a cavity to the base of a similar model; the RSB mode was weakened as the depth of the cavity (d) was increased, until saturation was reached for d > 0.24 H. A smaller improvement (around 2% over the uncontrolled case) has been seen from the feedback control system developed by Brackston et al. [5].

Grandemange et al. [12] also found that the RSB mode disappears when the horizontal symmetry of the time averaged wake is broken by perturbing one of the two horizontal shear layers with a control cylinder placed along the same direction, yielding noticeable variations in the lift over the uncontrolled case. A similar trend was noticed by Barros [2] studying the sensitivity of the wake past a square-back Ahmed body to steady perturbations applied underneath the model. He noticed that a perfect bi-modal behaviour along the transverse direction can be seen only when the steady state wake topology shows a good level of symmetry along vertical direction. The RSB mode, in fact, tends to weaken until disappearing as the horizontal symmetry is progressively lost. This agrees with the behaviour found by Perry et al. [18] looking at the effects produced by high aspect ratio tapers applied to the top and bottom trailing edges of a Windsor body, with the aim of replicating the shape optimisation process that is common for many modern production vehicles. It was seen that the tendency of the wake to develop a bi-stable behaviour tends to decrease as the distance between the top and bottom shear layers is reduced by the upwash or downwash induced by the small slants. However, a reduction of the drag generated by the rear facing surfaces $C_{D_{Rear}}$ was obtained only when the weakening of the RSB mode was accompanied with the re-symmetrisation of the time averaged wake topology along the vertical direction. An even bigger improvement of $C_{D_{Rear}}$ was achieved by Pavia et al. [16] forcing the lateral symmetry of the wake by applying similar tapers to the vertical trailing edges of the same model, leaving the horizontal edges squared.

The RSB mode and its effects on the base drag as well as the lateral force experienced by the model are further investigated in the present work by considering a geometry that, although being still fairly simple, is more representative of a real vehicle than those considered for example in Grandemange et al. [11] and Perry et al. [18], thanks to the presence of four rotating wheels.

2 Experimental Methodology

2.1 The Windsor Body

A Windsor body similar to that used in the work of Perry et al. [18] has been considered. The model is a simplified geometry, approximately equivalent to a 1/4 scale passenger car (Length L = 1044 mm, width W = 389 mm, height H = 289 mm), resulting in a tunnel blockage of 4.4%.

In the case considered in the present work, the model has been equipped with four rotating aluminium wheels, machined from solid. Each wheel has an external diameter D_w of 150 mm and a maximum width of W_w of 55 mm. The front axle is placed 280 mm upstream of the centre of the model while the rear axle is located 357 mm downstream of the same point, resulting in a wheelbase l of 637 mm, with the front and the rear overhangs measuring respectively l_f = 242 mm and l_r = 165 mm. The torque needed to spin the wheels was provided by four electrical brushless motors, with an external diameter D_m of 60 mm and a maximum width of W_m of 48 mm, mounted inside each wheel. The wheels were housed inside four 3D printed wheel arches, with a diameter D_{wa} of 170 mm and a depth of 96 mm at the front and 76 mm at the rear. When the wheels were not in use, the wheel arches were replaced with four 3D printed covers mounted flush with the side and underbody surfaces.

The model is also equipped with a removable rear section that allows testing of multiple rear taper configurations. Each taper has a chord c of 45 mm. Besides the pure square-back configuration, four cases have been studied, based on the results of Pavia et al. [16]: in the first two cases a slant with an angle of 12° was applied to either the upper or lower horizontal trailing edge, leaving the opposite edge squared, whilst two more tests were performed applying the same slant to both the horizontal or the vertical trailing edges. For the sake of clarity, ϕ_t denotes the angle formed by the top taper with the flat section of model roof, ϕ_b represents the angle between the bottom chamfer and the flat model underbody, whereas ϕ_s indicates the angle formed between each vertical taper and flat lateral surface of the model (see Fig. 1a).

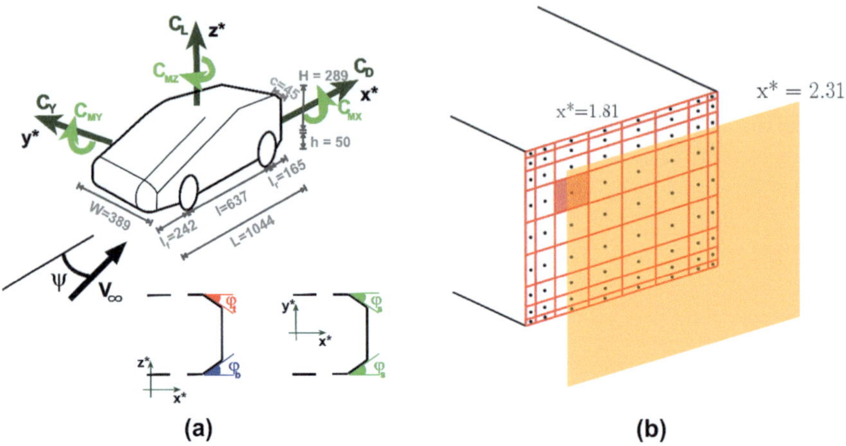

Fig. 1. a schematic representation of the Windsor body; b representation of the model base with the pressure tappings and the 2D-3C PIV plane considered in the present work. The red square on the model base sketched in b denotes the area associated with the nth tap used for the estimation of the area weighted drag. All the dimensions reported in a are expressed in mm. For the reference system see SAE [20].

The model was mounted via four pins (M8 threaded bar) to the six component balance located beneath the working section floor. The pins were placed between the two axles, 27.5 mm inboard of the model sides. The wheels were mounted separately from the model, bolted to the balance turntable on purpose-designed steel brackets.

For all the tests the ground clearance was set at 50 mm (17.3% of the model height), with a tolerance of 0.2° for the pitch angle. The contact between the wheels and the ground was represented with a 3 mm deep machined scallop underneath each wheel.

The SAE coordinate system (SAE [20]) is used throughout; the X axis is aligned with the flow in the downstream direction, the Z axis is vertical, positive upwards, and the Y axis follows a right handed coordinate system. The origin is on the ground plane at mid wheelbase, mid track (Fig. 1a). All the quantities presented throughout the paper have been normalised using the model height H as the reference length and the value of the free stream velocity V_∞ and are denoted with the superscript *.

2.2 The Wind Tunnel

All experiments were carried out in the Loughborough University Large Wind Tunnel. This test facility, described in Johl [13], has a working section measuring

1.92 × 1.32 × 3.6 m ($W_T \times H_T \times L_T$) and is equipped with a fixed floor without any upstream boundary layer treatment. In empty conditions, the free stream turbulence level inside the test section is approximately 0.2%, with a flow uniformity of ± 0.4% of the mean flow value. In this state, the boundary layer thickness at the model origin was measured to be equal to δ_{99} = 64 mm. All tests were performed with a free stream velocity of 40 m/s, corresponding to a Reynolds number Re_H of 7.7 × 10^5 based on the model height. When spinning, the rotational speed of the wheels was set manually to match the tunnel free stream velocity, with an accuracy of ±50 RPM.

2.3 Balance Measurements

The aerodynamic loads acting on the main body were recorded by means of an Aerotech® six-component virtual centre balance, located under the working section of the tunnel. The balance features analogue to digital conversion at the load cell to minimise signal degradation, and an automated yaw mechanism with a positional accuracy of 0.1°. Further information can be found in Johl [13]. The aerodynamic loads were sampled at 100 Hz for 630 s. Prior to logging the data, a 30 s settling time was used for all measurements. The recorded values of the forces have been normalised using Eq. 1:

$$C_{Fi} = \frac{Force}{0.5\,\rho S V_{\infty}^{2}} \tag{1}$$

where ρ is the air density and S is the projected model frontal area (S = 0.1124 m^2). The coefficients were corrected for blockage effects using Eq. 2:

$$C_{Fi_{Corr}} = C_{Fi}(1 - B)^2 \tag{2}$$

where B denotes the blockage value, given by the ratio between the model frontal area S and the tunnel working section cross sectional area S_T.

For the sake of simplicity, only the results related to the side force C_Y are presented in this work as this is the component of the aerodynamic force most affected by the action of the RSB mode (Grandemange et al. [9] and Grandemange et al. [12]).

2.4 Pressure Measurements

The pressure acting on the model base and tapers was recorded by populating the rear facing surfaces with a grid of pressure taps connected via flexible tubes to a pair of Chell® CANdaq miniature pressure scanners (with a manufacturer quoted accuracy of ±1.47 Pa) mounted inside the model. Each scanner is equipped with 64 piezo-resistive pressure sensors, paired to temperature sensors to allow the correction of inaccuracies introduced by temperature drift. Between 72 and 126 taps were used, depending on the configuration. The taps were placed with a finer distribution close to the model edges, to get a more accurate representation of the pressure distribution in the regions with the highest gradients (see Fig. 1b). Data was recorded at 260 Hz for 630 s. The free-stream dynamic and static pressures were acquired 1870 mm upstream of the model, at the start of the working section. Signal was corrected for distortion applying the correction proposed by (Sims-Williams and Dominy [21]). Once the pressure coefficients were calculated, Eq. 3, the results were corrected for blockage using the MIRA correction (based on continuity), Eq. 4:

$$C_p = \frac{p - p_\infty}{0.5\,\rho S V_\infty^2} \tag{3}$$

$$1 - C_{pCorr} = \left(1 - C_p\right)(1 - B)^2 \tag{4}$$

where p_∞ is the freestream static pressure. The contribution of the model rear facing surfaces to the aerodynamic force experienced by the model was estimated by integrating the measured pressure field:

$$C_{DRear} = -\frac{1}{S}\int_S C_p \cdot dS \simeq -\frac{1}{S}\sum_{i=1}^{N_{tap}} C_{pi} S_i, \tag{5}$$

where C_{pi} is the time averaged value of the pressure coefficient recorded by the nth tap and S_i is the projection of the associated area (see Fig. 1b).

2.5 PIV Measurements

Stereoscopic particle image velocimetry (PIV) measurements were performed considering a plane orthogonal to the free stream velocity, at $x^* = 2.31$ from the origin of the reference system (see Fig. 1b). Two LaVision® CMOS cameras with a resolution of 2560 × 2160 pixels and a pixel size of 6.5 × 6.5 μm^2 were employed, together with a Litron® 200 mJ double pulsed Nd:YAG laser and PIVTec 45 atomiser, generating 1 μm droplets of DEHS. The cameras were mounted on a pair of aluminium rails placed inside the tunnel working section, following the arrangement used by Wood [25]. The separation angle between the cameras was ≈50°. Each sensor was equipped with a Nikon® lens, with a focal length of 50 mm, mounted on a tilt system in order to satisfy the Scheimplfug criterion (Prasad [19]). The aperture was set at $f_\# = 4$. The resultant field of view ($W_f \times H_f$) was 520 × 400 mm.

Data acquisition was performed at 15 Hz. During each acquisition, 2000 statistically independent image pairs were captured, matching the total sampling time used by Perry et al. [18] (133.3 s). The data was post-processed using DaVis 8.2.0. Image pre-processing was applied to the recordings in order to mitigate the effects of image distortion, background light intensity and spurious reflections. A calibration correction based on a pinhole model was applied, background subtraction was performed and the outmost regions of the field of view were discarded using a geometric mask. The size of the field of view was therefore reduced to 430 × 340 mm. A multi-pass scheme for cross-correlation was then applied (Willert and Gharib [24]) starting with an interrogation window with a size of 64 × 64 pixels and a 50% overlap between cells and ending with windows of 32 × 32 mm pixels and a 75% overlap. This resulted in a spatial resolution of 2.8×10^{-6} m^2. The level of uncertainty associated with the measurements was estimated to be 0.5% and 1.5% of the mean and RMS values of the free stream velocity with a 99% confidence level (Benedict and Gould [3]).

3 Results and Discussion

3.1 Square-Back Configuration: Effects of the Wheels on the RSB Mode

A first series of tests was performed considering the same square-back configuration without wheels studied in Perry et al. [18]. Stereoscopic PIV was performed inside the near wake, 145 mm downstream of the model base ($x^* = 2.31$ from the origin of the reference system defined in Fig. 1a). The time averaged results, reported in Fig. 2a, show a good level of symmetry along the vertical direction but a slight asymmetry between the two lateral sides of the model, in agreement with that shown by the horizontal 2D PIV presented in Perry et al. [18]. Two reflectional symmetry breaking (RSB) states were then extracted, dubbed *L state* and *R state* (Fig. 2b and c), by applying the same conditional average technique used in Perry et al. [18]. Each of these two states is characterised by the presence of two counter rotating vortices aligned with one of the two vertical edges, originating a low pressure region that is responsible of the creation of the horizontal pressure gradient already documented in other works (Evrard et al. [7] and Grandemange [8]). The bi-stable mode consists of a random switch between these two reflectional symmetry breaking states. A good level of lateral symmetry in the time averaged wake topology can be obtained only when the acquisition time is long enough to capture the two states for approximately the same number of realisations.

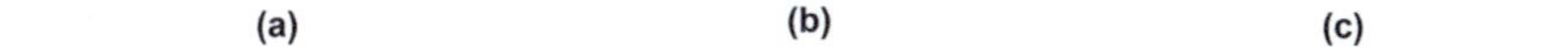

Fig. 2. Square-back configuration without wheels: a time averaged wake topology for $x^* = 2.31$; b *L state* and c *R state* extracted following the same method used in Perry et al. [18].

Due to the action of this mode, a high level of unsteadiness can be seen close to the core of the wake itself, as shown by the values of the normal components of the Reynolds stresses reported in Fig. 3. In particular, two peaks can be seen in the contour plot obtained for the streamwise component T^*_{uu} (Fig. 3a), corresponding with the centre of each one of the vertical halves of the model base. This matches the distribution of the base pressure fluctuations reported in Pavia et al. [16] for the same model. The level of fluctuations in these regions is comparable with that seen in the shear layers surrounding the wake. A similar intensity is also seen in the lateral component of the Reynolds stresses T^*_{vv} recorded in the same area (Fig. 3b), although this time the peaks are centred with the base and values tend to spread along the horizontal direction rather than being clustered around two lobes. No activity, on the other hand, can be seen in the vertical component T^*_{ww} (Fig. 3c), apart from the regions corresponding to the horizontal and vertical shear layers. All these elements suggest that the RSB mode involves a lateral as well as longitudinal motion of the wake, with the latter being probably related to the pumping motion described by Duell and George [6] and Volpe et al. [23].

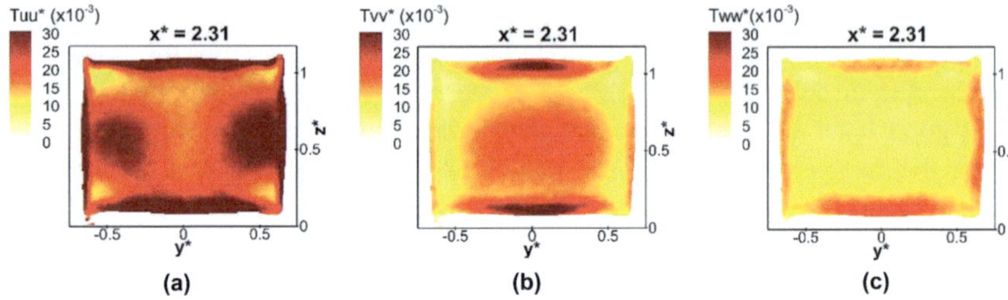

Fig. 3. Normal components of the Reynolds stress tensor for the square-back model without wheels: a $T^*_{uu} = \langle u^{*\prime 2} \rangle$; b $T^*_{vv} = \langle v^{*\prime 2} \rangle$; c $T^*_{ww} = \langle w^{*\prime 2} \rangle$.

The effects produced by the presence of the wheels on the RSB mode as well as the base drag were studied by applying the four wheels described in Sect. 2.1 to the square-back configuration.

The addition of just the front axle yields a slight decrease in the values of pressure recorded around the lower portion of the base over the square-back configuration tested in Pavia et al. [16] (Fig. 4a). The increase in the suction in this area is counteracted by a better pressure recovery over the upper half of the base, yielding only a marginal variation in the base drag ($\Delta C_{DRear} \simeq$+1 count). No differences can be noticed between the stationary and the rotating case. The presence of a strong bi-stable motion is still recognisable, as seen from the distribution of the pressure fluctuations over the model base that is in good agreement with the two-lobe distribution shown for example by Volpe et al. [23] for a bi-stable case (Fig. 5a).

A noticeable change in the distribution of the pressure fluctuations can be seen when the rear wheels are added to the model (Fig. 5b and c). In this case, the region with the

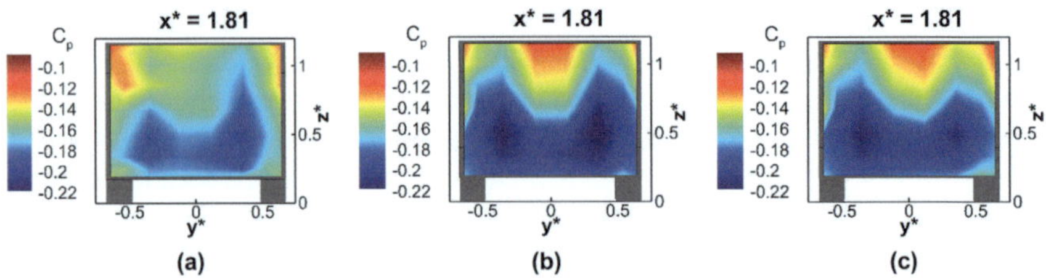

Fig. 4. Time averaged base pressure distribution for the square-back configuration: a front axle only (with stationary wheels); b four stationary wheels; c four rotating wheels.

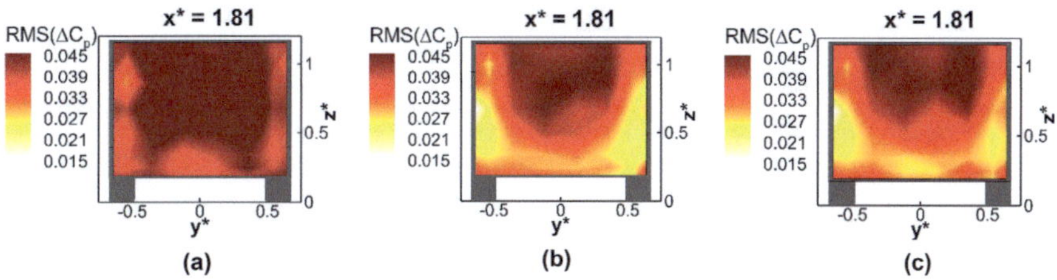

Fig. 5. Base pressure fluctuations for the square-back configuration: a front axle only (with stationary wheels); b four stationary wheels; c four rotating wheels.

highest level of fluctuations is reduced in size and moved towards the upper edge of the base, in a similar fashion to that reported by Perry et al. [18] when the upwash in the near wake was increased by adding a small taper to the bottom trailing edge of the model. On this occasion, however, the two-lobe structure becomes harder to see inside the region with the highest values of $\text{RMS}(\Delta C_p)$, suggesting an even further weakening of the RSB mode. No significant differences can be seen between the stationary (Fig. 5b) and the rotating (Fig. 5c) case. Similar conclusions hold when comparing the topologies of the time averaged base pressure distributions for the same configurations (Fig. 4b and c). In both cases a further growth of the suction previously noticed in the bottom half of the base can be seen. This results in an increase of C_{DRear} between 13 counts and 7 counts over the square-back configuration without wheels, with the rotating configuration showing a better pressure recovery than the stationary case.

The origin of this low pressure region can be inferred by looking at the time averaged velocity field presented in Fig. 6a. The PIV data shows the presence of a pair of counter rotating vortices acting in the gap between the two rear wheels. These vortices are quite stable, as evidenced by the almost null values of the three normal components of the Reynolds stresses in this region (Fig. 7). This is in contrast to the higher level of instability seen in the pair of recirculating structures acting on the upper portion of the base, although the values obtained for T^*_{uu} and T^*_{vv} are somehow lower than in the previous case (Fig. 3). All these elements suggest the existence of a stabilising effect on the wake produced by the vortical structures created by the wheels. The action of these vortices appears to counteract the tendency of the upper recirculating structures to align their centres alternately with the two vertical trailing edges,

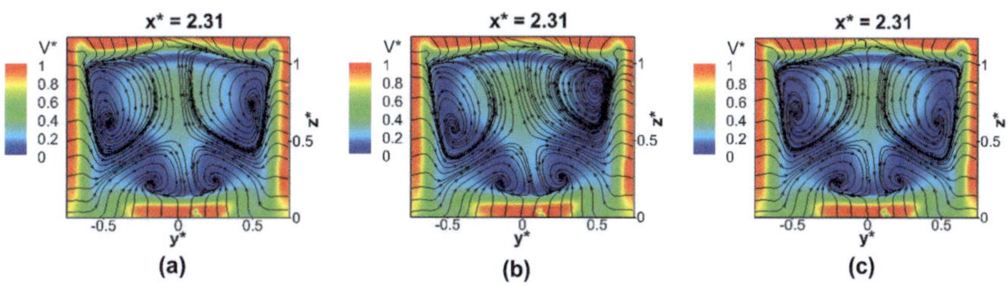

Fig. 6. Square-back configuration with stationary wheels: a time averaged wake topology for $x^* = 2.31$; b *L state* and c *R state* extracted by applying the same method used in Perry et al. [18].

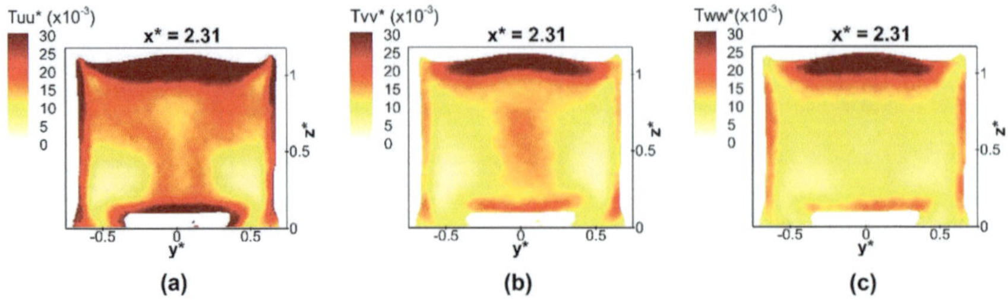

Fig. 7. Normal components of the Reynolds stress tensor for the square-back model with stationary wheels: a $T^*_{uu} = \langle u^{*\prime 2} \rangle$; b $T^*_{vv} = \langle v^{*\prime 2} \rangle$; c $T^*_{ww} = \langle w^{*\prime 2} \rangle$.

resulting in the RSB states previously described. As a consequence, the lateral "displacement" of the wake is limited and the only difference between the two states, when they are isolated using the method proposed in Perry et al. [18], is in the direction of the reverse flow impinging on the upper portion of the base (Fig. 6b and c).

3.2 Effects Produced by Horizontal Trailing Edge Tapers on the RSB Mode

The sensitivity of the wake to small changes in the shape of the model trailing edges was studied by applying small tapered surfaces, with a slant angle of 12°, to the edge of the base. Three configurations are reported: top slant only, bottom only and both top and bottom tapers. All tests are with wheels in place, as already described in Sect. 2. For the sake of simplicity only the results related to the stationary case are presented, since having the wheels rotating did not produce noticeable changes in either the unsteady behaviour of the wake or its time averaged topology, as described in Sect. 3.1.

The results related to the time averaged base pressure distribution and wake topology are presented in Figs. 8 and 9. It is shown that a good horizontal symmetry in the distribution of the mean values of C_p is obtained only when the 12° taper is applied to just the top edge of the base (Fig. 8a). In this case a good balance is reached between the upper and lower pairs of vortices, as the strength of the upper recirculating structures is reduced by the action of the pair of counter rotating trailing vortices forming at the edges of the slanted surface (Fig. 9a) and their cores are moved

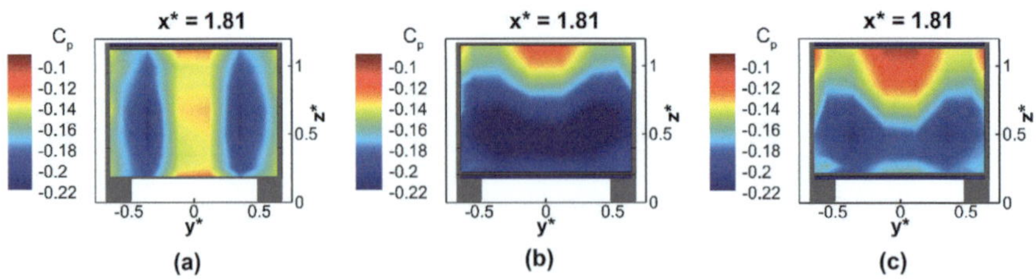

Fig. 8. Time averaged base pressure distribution for different rear-end configurations with stationary wheels: a $\phi_t = 12°, \phi_b = 0°$; b $\phi_t = 0°, \phi_b = 12°$; c $\phi_t = 12°, \phi_b = 12°$.

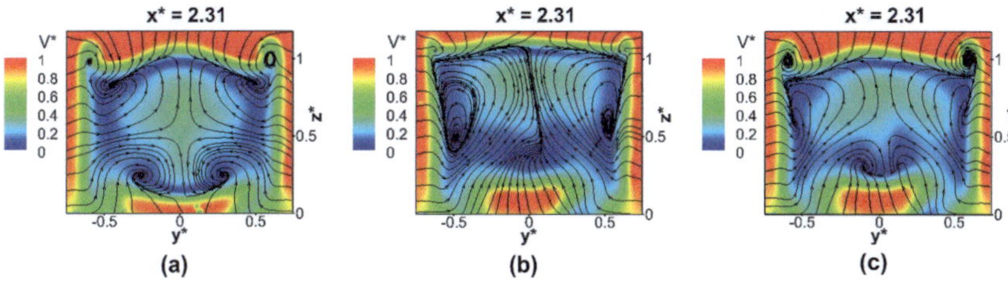

Fig. 9. Time averaged velocity field at $x^* = 2.31$ for different rear-end configurations with stationary wheels: a $\phi_t = 12°, \phi_b = 0°$; b $\phi_t = 0°, \phi_b = 12°$; c $\phi_t = 12°, \phi_b = 12°$.

outwards allowing the formation of a "high" pressure region around the centre of the base. A better pressure recovery is therefore achieved, as highlighted by the decrease of 8 counts in the values of C_{DRear}(−12 counts for the rotating case) over the square-back configuration with stationary wheels (although C_{DRear} is still 5 counts higher than in the case without wheels). The trailing vortices seem to further contribute to the stabilisation of wake, as suggested by the higher degree of lateral symmetry shown by the time averaged velocity field as well as the shrinking of the regions showing the highest values of the longitudinal and lateral components of the Reynolds stress (Fig. 10). Nevertheless, high values of T^*_{uu} and T^*_{vv} can still be seen corresponding to the inner region of the wake, matching the peak in the values of $RMS(\Delta C_p)$ visible in Fig. 11a. This suggests that a certain degree of "motion" of the reverse flow impinging on the base is still present, although it is more limited than in the previous cases due to the constraints imposed by the combined action of the vortical structures generated by the wheels and the top slant.

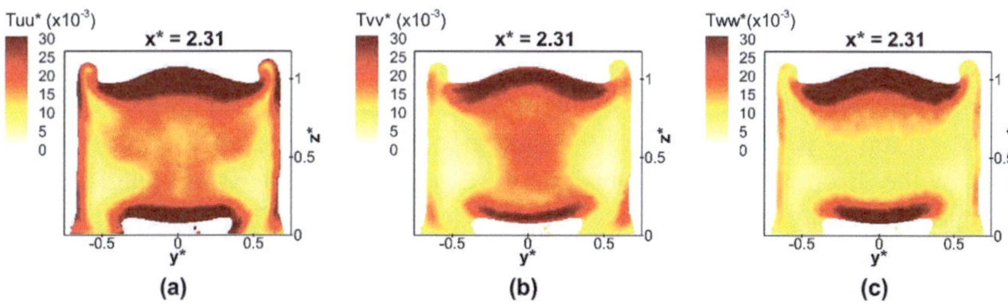

Fig. 10. Normal components of the Reynolds stress tensor for the configuration with $\phi_t = 12°$, $\phi_b = 0°$ and stationary wheels: a $T^*_{uu} = \langle u^{*\prime 2} \rangle$; b $T^*_{vv} = \langle v^{*\prime 2} \rangle$; c $T^*_{ww} = \langle w^{*\prime 2} \rangle$.

An opposite effect in terms of base pressure distribution is obtained when the 12° taper is applied only to the bottom trailing edge. In this case, the wake is dominated by the upwash generated by the combined action of the wheels and the diffuser (Fig. 9b). As a result, just a pair of counter rotating vortices can be seen in the wake, extending for almost the entire height of the base. This leads to a further growth in the size of the low pressure region previously described for the square-back case with stationary

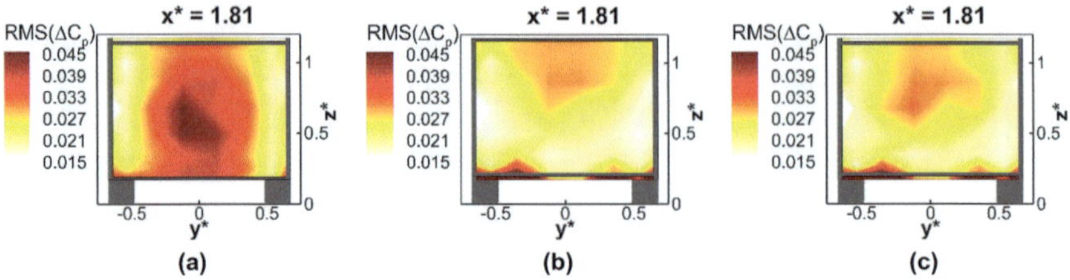

Fig. 11. Base pressure fluctuations for different rear-end configurations with stationary wheels: a $\phi_t = 12°, \phi_b = 0°$; b $\phi_t = 0°, \phi_b = 12°$; c $\phi_t = 12°, \phi_b = 12°$.

wheels (Fig. 8b), resulting in a 15 count increase in the value of the area weighted pressure drag over the same configuration (+8 counts for the rotating case). At the same time, a significant drop can be seen in the fluctuations of the pressure data, suggesting the disappearance of the bi-stable mode (Fig. 11b).

A similar trend can be seen from the contour plot of RMS(ΔC_p) obtained for the case with $\phi_t = 12°$ and $\phi_b = 12°$ (Fig. 11c). However, the action of the bottom diffuser is largely counteracted by the top slant. Therefore, the results obtained for the time averaged velocity field (Fig. 9c) and the related base pressure distribution (Fig. 8c) are somewhere in between those obtained in the previous two cases. This is further confirmed by the values obtained for C_{DRear} which are just 1 count higher than those obtained for the configuration with just the top slant in place, for both the stationary and the rotating case.

3.3 Effects Produced by Vertical Trailing Edge Tapers on the RSB Mode

When 12° tapers are applied to the vertical trailing edges of the model, leaving the horizontal edges squared, a strengthening of the RSB mode can be noticed. In this case, the size of the vortex pair acting in the lower portion of the base is reduced (Fig. 12a) because the pressure difference between the lateral surfaces and the underbody is lowered by the suction introduced by the tapers. As a consequence, the size of the low pressure region is reduced (Fig. 13a), leading to a better pressure recovery, as highlighted by the 21 count reduction obtained for C_{DRear} compared to the square-back case with stationary wheels (−23 counts for the rotating case). The weakening of these

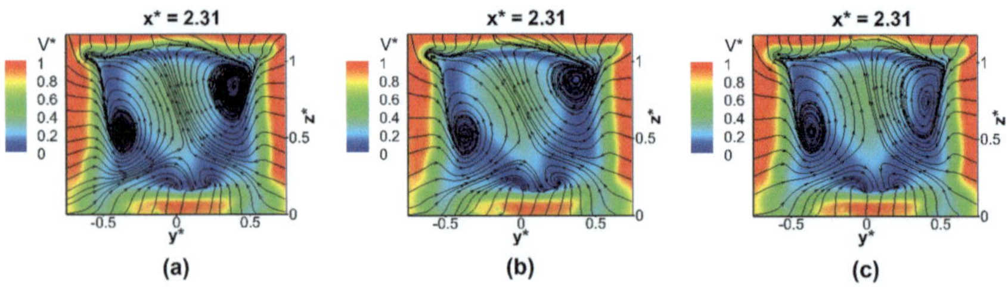

Fig. 12. Configuration with $\phi_s = 12°$ and stationary wheels: a time averaged wake topology for $x^* = 2.31$; b RSB state and c RSP state extracted by applying the same method used in Perry et al. [18].

vortices allows for a higher degree of movement of the upper structures, which results in an increase of the wake unsteadiness seen in both the normal components of the Reynolds stress tensor in the near wake (Fig. 14) and the pressure fluctuations recorded on the base (Fig. 13b). As the wake becomes more unsteady, its sensitivity to any imperfection in the experimental setup is amplified (Evrard et al. [7]), explaining the clear asymmetry that can be found in the time averaged results obtained from both the PIV data (Fig. 12a) and the base pressure data (Fig. 13a). In this case the wake seems to switch between a (left) lateral symmetry breaking (RSB) state (Fig. 12b) and a reflectional symmetry preserving (RSP) state (Fig. 12c), in such a way that the averaged results are clearly biased towards one side.

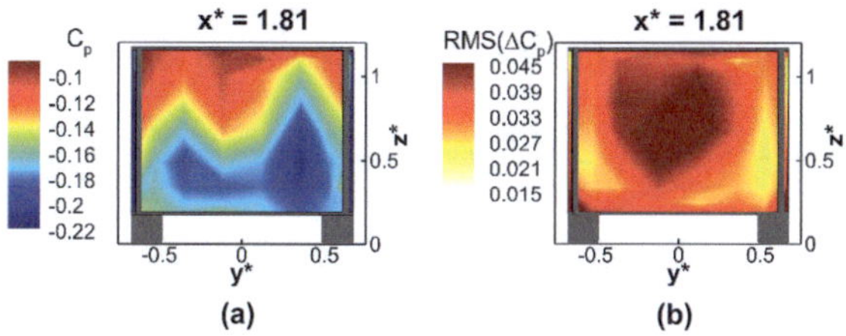

Fig. 13. Configuration with $\phi_s = 12°$ and stationary wheels: a time averaged base pressure distribution; b pressure fluctuations over the model base.

3.4 Effects of the RSB Mode on the Side Force

In order to assess the effects of the RSB mode on the lateral component of the aerodynamic force, an approach similar to that used in Perry et al. [18] was followed, considering the probability density function distribution (PDF) distribution of

the values recorded for C_Y after having resampled the original signal at 5 Hz. The results are presented in Fig. 15.

For the square-back configuration, a clear bi-modal distribution can be seen for the values of C_Y when just the front axle is added to the model. No big differences indeed can be seen between this case and the model without wheels tested in Perry et al. [18], confirming the conclusions in Sect. 3.1. When all four wheels are added to the geometry, the two peaks seen in the first case disappear, following the weakening of the

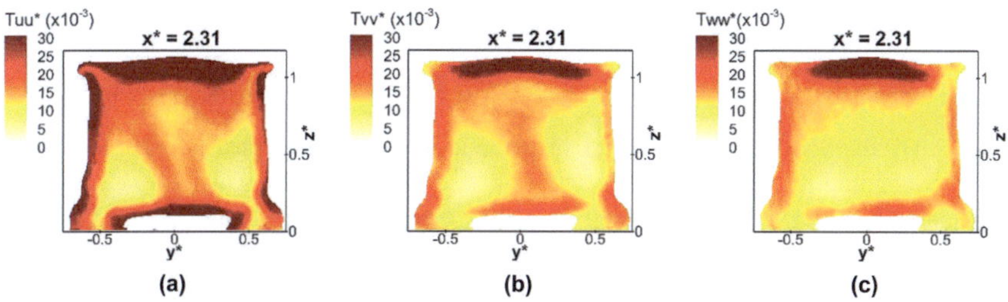

Fig. 14. Normal components of the Reynolds stress tensor for the model with $\phi_s = 12°$ and stationary wheels: a $T^*_{uu} = \langle u^{*\prime 2} \rangle$; b $T^*_{vv} = \langle v^{*\prime 2} \rangle$; c $T^*_{ww} = \langle w^{*\prime 2} \rangle$.

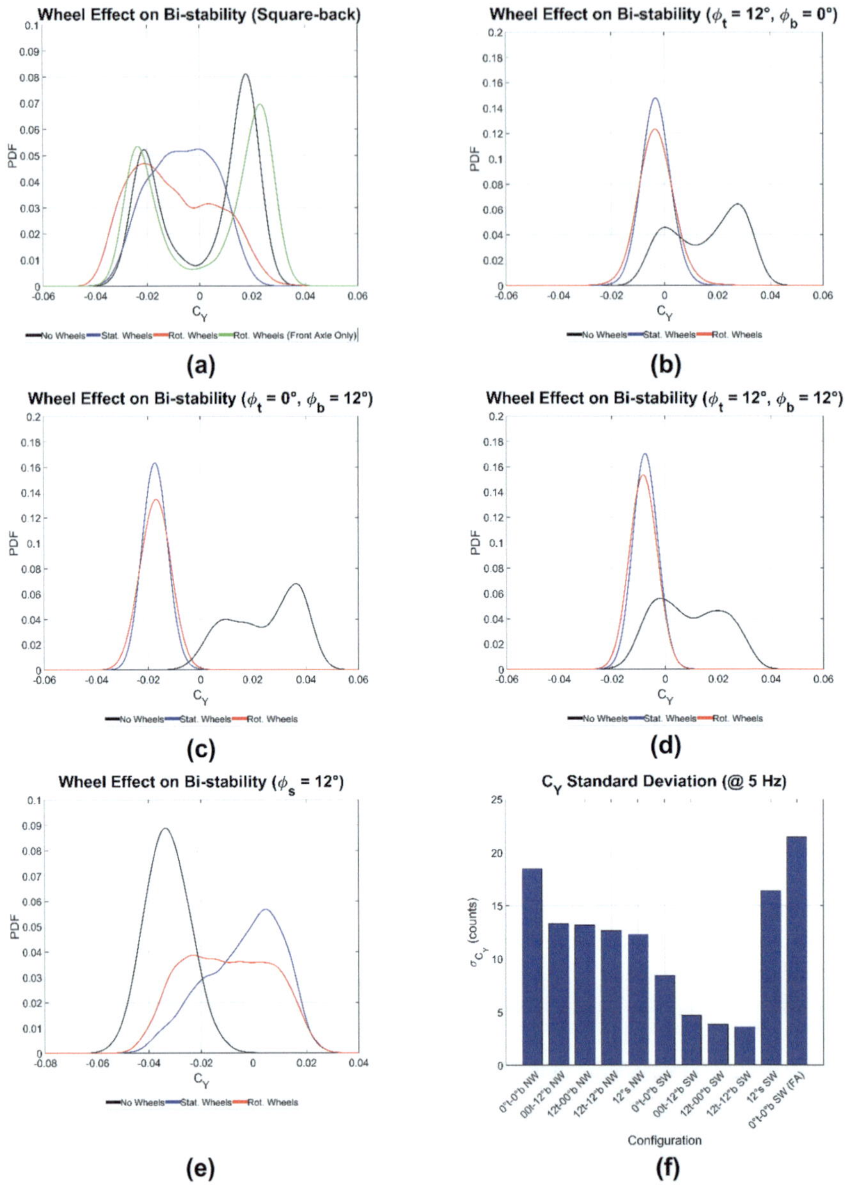

Fig. 15. PDF distributions of the values of the side force C_Y resampled at 5 Hz for different rear-end configurations without wheels and with stationary or rotating wheels: a square-back configuration; b model with $\phi_t = 12°$ and $\phi_b = 0°$; c model with $\phi_t = 0°$ and $\phi_b = 12°$; d model with $\phi_t = 12°$ and $\phi_b = 12°$; e model with $\phi_s = 12°$. For each configuration C_Y has been recorded at the value of the yaw angle Ψ corresponding to the most symmetric base pressure distribution along the lateral direction. f values of the standard deviation of C_Y recorded for the same configurations either with stationary wheels (SW) or without wheels (NW).

RSB mode. A high level of dispersion, however, can still be seen for both the stationary and rotating case (Fig. 15a).

The dispersion in the values of C_Y is noticeably reduced when the 12° taper is applied to either the top (Fig. 15b), the bottom (Fig. 15c) or both (Fig. 15d) the horizontal trailing edges. For all these configurations a normal distribution is obtained for the recorded values of the lateral force, in contrast with that seen from the data recorded using the same configurations without wheels. As in other cases no differences appear between the stationary and rotating case.

An opposite trend appears when the tapers are applied to the vertical trailing edges. In this case the level of dispersion associated with the values recorded for C_Y increases when adding wheels, further confirming that already stated in Sect. 3.3 for the same configuration.

Similar conclusions can be drawn by looking at the values of the standard deviation σ_{CY} associated to the same signals (Fig. 15f), as suggested by Grandemange et al. [9]. Indeed, the highest values of σ_{CY} are associated with the most bi-stable configurations (namely the square-back without wheels or with just the front axle in place), whilst all the configurations tested with wheels and at least one horizontal taper in place show the lowest level of uncertainty because the effects produced by the RSB mode are much more limited.

4 Summary and Conclusions

In this paper the effects produced by wheels on the base drag and bi-stable RSB mode have been investigated by means of pressure tappings, 2D - 3C PIV and balance measurements. The Windsor body used Perry et al. [18] was considered.

For the square-back configuration, the results showed a general increase in the drag associated with the rear facing surfaces, mainly due to the strengthening of the suction over the base as a consequence of the formation of a pair of counter rotating vortices acting close to the lower portion of the base associated with the gap between the rear wheels. These structures have been shown to counteract the tendency of the upper pair of vortices to alternately align their cores with the model vertical trailing edges, resulting in the bi-stable RSB mode described by Grandemange et al. [11].

The sensitivity of the RSB mode to small changes in the shape of the trailing edges was assessed by applying short high aspect ratio tapers, with a slant angle of 12°, to either the horizontal or the vertical edges of the base. The results showed that, when the slant is applied to the bottom trailing edge, the RSB mode disappears due to the stabilising effect of the upwash generated by the combined action of the wheels and tapered surface. As a consequence, the PDF distribution of the values recorded for the side force switches from a bi-modal distribution to a normal distribution, accompanied with a noticeable reduction of the standard deviation associated with the same data.

A strengthening of the bi-stable mode was seen when the tapers were applied to the side trailing edges, in contrast with the results reported by Pavia et al. [16] for the same model without wheels. At the same time, an improvement in the pressure recovery over the base was noticed.

In all the cases, no significant differences were seen in either the wake topology or the wake dynamics between having the wheels stationary or rotating, apart from a decrease in the values of the drag generated by the rear facing surfaces in the latter case.

Acknowledgements. The authors would like to thank Jaguar Land Rover for the financial support. Thanks are also due to Mr. David Cooper and Mr. Nigel Lines for their excellent work in manufacturing the models and keeping the test facility always in optimal conditions.

References

1. Ahmed, S., Ramm, G., and Faitin, G.: Some salient features of the time-averaged ground vehicle wake. Technical report, Society of Automotive Engineers, Inc., Warrendale, PA (1984)
2. Barros, D.: Wake and drag manipulation of a bluff body using fluidic forcing. Ph.D. thesis, École Nationale Supérieure de Mécanique et de Aérotechnique (ENSMA) (2015)
3. Benedict, L., Gould, R.: Towards better uncertainty estimates for turbulence statistics. Exp. Fluids **22**(2), 129–136 (1996)
4. Bonnavion, G., Cadot, O., Évrard, A., Herbert, V., Parpais, S., Vigneron, R., Délery, J.: On multistabilities of real car's wake. J. Wind Eng. Ind. Aerodyn. **164**, 22–33 (2017)
5. Brackston, R., de la Cruz, J.G., Wynn, A., Rigas, G., Morrison, J.: Stochastic modelling and feedback control of bistability in a turbulent bluff body wake. J. Fluid Mech. **802**, 726–749 (2016)
6. Duell, E.G., George, A.: Experimental study of a ground vehicle body unsteady near wake. Technical report, SAE technical paper (1999)
7. Evrard, A., Cadot, O., Herbert, V., Ricot, D., Vigneron, R., Délery, J.: Fluid force and symmetry breaking modes of a 3d bluff body with a base cavity. J. Fluids Struct. **61**, 99–114 (2016)
8. Grandemange, M.: Analysis and control of three-dimensional turbulent wakes: from axisymmetric bodies to road vehicles. Ph.D. thesis, Palaiseau, Ecole polytechnique (2013)
9. Grandemange, M., Cadot, O., Courbois, A., Herbert, V., Ricot, D., Ruiz, T., Vigneron, R.: A study of wake effects on the drag of Ahmed's squareback model at the industrial scale. J. Wind Eng. Ind. Aerodyn. **145**, 282–291 (2015)
10. Grandemange, M., Gohlke, M., Cadot, O.: Bi-stability in the turbulent wake past parallelepiped bodies with various aspect ratios and wall effects. Phys. Fluids (1994-Present) **25**(9), 95–103 (2013)
11. Grandemange, M., Gohlke, M., Cadot, O.: Turbulent wake past a three-dimensional blunt body. Part 1: global modes and bi-stability. J. Fluid Mech. **722**, 51–84 (2013)
12. Grandemange, M., Gohlke, M., Cadot, O.: Turbulent wake past a three-dimensional blunt body. Part 2: experimental sensitivity analysis. J. Fluid Mech. **752**, 439–461 (2014)
13. Johl, G.: The design and performance of a 1.9 m $\times$ 1.3 m indraft wind tunnel. Ph.D. thesis, © Guru Johl (2010)
14. Krajnovic, S., Davidson, L.: Numerical study of the flow around a bus-shaped body. J. Fluids Eng. **125**(3), 500–509 (2003)
15. Makihara, T., Kitamura, T., Yamashita, T., Maeda, K., Kato, C., Takayama, T., Yamamoto, K., Yamade, Y., Suzuki, Y.: Identification of vortical structure that drastically worsens aerodynamic drag on a 2-box vehicle using large-scale simulations. SAE Int. J. Passeng. Cars-Mech. Syst. **9**, 592–602 (2016). 2016-01-1585

16. Pavia, G., Passmore, M., Gaylard, A.: Influence of short rear end tapers on the unsteady base pressure of a simplified ground vehicle. Technical report, SAE technical paper (2016)
17. Perry, A.-K., Almond, M., Passmore, M., Littlewood, R.: The study of a bi-stable wake region of a generic squareback vehicle using tomographic piv. SAE Int. J. Passeng. Cars-Mech. Syst. **9** ,743–754 (2016). 2016-01-1610
18. Perry, A.-K., Pavia, G., Passmore, M.: Influence of short rear end tapers on the wake of a simplified square-back vehicle: wake topology and rear drag. Exp. Fluids **57**(11), 169 (2016)
19. Prasad, A.K.: Stereoscopic particle image velocimetry. Exp. Fluids **29**(2), 103–116 (2000)
20. SAE: Surface vehicle recommended practice. Technical report J1594. SAE International (2010)
21. Sims-Williams, D.B., Dominy, R.: Experimental investigation into unsteadiness and instability in passenger car aerodynamics. Technical report, SAE technical paper (1998)
22. Varon, E., Eulalie, Y., Edwige, S., Gilotte, P., Aider, J.-L.: Chaotic dynamics of large-scale structures in a turbulent wake. Phys. Rev. Fluids **2**(3), 034604 (2017)
23. Volpe, R., Devinant, P., Kourta, A.: Experimental characterization of the unsteady natural wake of the full-scale square back Ahmed body: flow bi-stability and spectral analysis. Exp. Fluids **56**(5), 1–22 (2015)
24. Willert, C.E., Gharib, M.: Digital particle image velocimetry. Exp. Fluids **10**(4), 181–193 (1991)
25. Wood, D.: The effect of rear geometry changes on the notchback flow field. Ph.D. thesis, ©Daniel Wood (2015)

Investigation of Time-Resolved Nozzle Interference Effects

Christoph Schoenleber(✉), T. Kuthada, Nils Widdecke, F. Wittmeier, and J. Wiedemann

FKFS/IVK, University of Stuttgart, Pfaffenwaldring 12, 70569 Stuttgart, Germany
Christoph.Schoenleber@fkfs.de

Abstract. Automotive wind tunnels still see an increasing significance in the vehicle development process. Nevertheless, the flow topology in the test sections of open jet wind tunnels is not yet understood completely. A large source for transient structures is the shear layer developing between the high velocity nozzle flow and the calmly air in the plenum.

For a better understanding, the shear layer is investigated in the symmetry plane with a multi-hole probe and PIV measurements in this paper. The analysis of the recorded data shows fluctuations with a Strouhal number of 0.46 based on the nozzle hydraulic diameter. This value matches other researches of jet flow. However, the concept of vortices separating at the nozzle edge seems to be incomplete. The PIV measurements reveal a wave structure in the shear layer at the nozzle exit. These waves release larger vortices moving along the outer boundary of the shear layer. Mode decomposition (POD) shows changing intensities of these vortices for varying wind tunnel velocities. Different flow structures are detected when delta wings are mounted to the nozzle edges.

1 Introduction

In recent years the accuracy in aerodynamic measuring and simulation increased. Regarding this development it is interesting to investigate wind tunnel interference effects in a time-resolved manner. This helps to further understand the effects and can lead to an improvement of the interference calculation methods.

Wind tunnel interference effects are part of automotive wind tunnel research since more than thirty years. The first methods to calculate interference, developed by Wiedemann and Mercker [1], are based on potential theory. Extended by a term for the horizontal pressure distribution [2], the impact of the procedure was improved. With the two-measurement method an accurate calculation of the influence of the wind tunnel pressure distribution on the vehicle's drag is possible [3]. Therefore the drag is measured with different wind tunnel pressure distributions. A different approach was chosen by Wickern and Schwartekopp [4]. They use the momentum balance and potential theory to calculate the nozzle gradient effects. The latest improvement was introduced by Hennig et al. [5] to further optimize the calculation of the collector effect.

Wind tunnel buffeting and transient effects in wind tunnels are object of many more investigations and not only a topic in automotive measurements. In case of automotive

J. Wiedemann (ed.), *Progress in Vehicle Aerodynamics and Thermal Management*,
https://doi.org/10.1007/978-3-319-67822-1_7

wind tunnels buffeting has a negative impact on flow quality and therefore on aerodynamic measurements as well as aeroacoustic investigations. An active system with large loud speakers to reduce fluctuations was presented by Wickern et al. [6]. A different measure is a compensation duct that connects the diffuser with the plenum. Its efficiency was verified by Beland [7] and it was integrated in several wind tunnels subsequently. Buffeting can moreover be reduced by plenum Helmholtz resonators [7] that can shift the plenum modes from resonance frequencies to less critical ones. As these measurements do not influence the fluctuations in the shear layer itself it is also possible to influence the natural vortices shed from the nozzle directly. Delta wings or Seiferth wings are installed in several wind tunnels [8], with the disadvantage of having a high background noise. This issue of high frequency noise could be reduced with the FKFS ***besst***® system [9].

Several generic investigations on jet flows were conducted in the last decades. However, the nozzles were much smaller and therefore the Reynolds numbers were in a range of up to 10^5. The model scale wind tunnel used in this investigation is typically operated at Reynolds numbers based on the hydraulic diameter of at least 5×10^6. Hence, the comparison of results between these cases is not recommended.

2 Fundamentals

In this section the used theory in fluid dynamics, wind tunnel effects and the measurement and simulation technique are recapitulated.

2.1 Fluid Mechanics

Although the most important formulae are well-known; they are pictured in the following passage. In addition good overviews about calculation methods for wind tunnel interference effects are given for example by Mercker [10].

As the investigation in this paper focuses on the nozzle interference effects, it is important to evaluate the shear layer developing between jet and plenum. Similar to boundary layers the displacement thickness and momentum thickness can be applied. In Schlichting [11] they are defined for steady-state incompressible flows as the displacement thickness

$$\delta_1 = \int_{y=0}^{\infty} \left(1 - \frac{u(y)}{U_\infty}\right) dy \tag{1}$$

and the momentum thickness

$$\delta_2 = \int_{y=0}^{\infty} \frac{u(y)}{U_\infty} \left(1 - \frac{u(y)}{U_\infty}\right) dy \tag{2}$$

with the local velocity u in x direction and the wind tunnel reference velocity U_∞. As a dimensionless magnitude the vorticity height is adopted from Haigermoser et al. [12]:

$$\delta_{\omega} = \frac{u_2 - u_1}{\left(\frac{\partial \bar{u}}{\partial y}\right)_{max}} \tag{3}$$

Here the velocities u_1 and u_2 are the plenum velocity (equals 0) and the reference velocity. An indicator for transient fluctuations, coherent or chaotic, is the turbulence intensity. It is calculated with the mean square of the velocity fluctuations in all three flow directions and averaged with the wind tunnel reference velocity.

$$Tu = \frac{1}{U_{\infty}} \cdot \sqrt{\frac{1}{3} \cdot \left(\overline{u'^2} + \overline{v'^2} + \overline{w'^2}\right)} \tag{4}$$

2.2 Wind Tunnel Effects

For a time-resolved study of the flow in the shear layer and test section different evaluations are used. Several resonance effects are known and investigated, for example by Beland [7], the most popular being the natural vortex shedding. The dimensionless frequency describing the vortex shedding is defined as the Strouhal number. It is calculated with a measured frequency f, the reference velocity and with a characteristic length L.

$$Sr = \frac{f \cdot L}{U_{\infty}} \tag{5}$$

For the characteristic length the hydraulic diameter of the nozzle D_h or the momentum thickness δ_2 are used. The typical value for the vortex shedding frequency found in literature for open-jet wind tunnels is $Sr_{Dh} \approx 0.34$. For automotive wind tunnels with a ¾-open-jet test section this value has to be increased by a factor of $\sqrt{2}$ to $Sr_{Dh} \approx 0.48$ for the fact that there is no shear layer on the floor and therefore no ring vortex. Instead a ¾-ring vortex develops. The value for the open jet is in good agreement with the step mode based on the height H given in Gong [13] with $Sr_H \approx 0.3 \ldots 0.4$, where two different shear layer modes are given to $Sr_{\delta 2} = 0.014$ and $Sr_{\delta 2} = 0.067$.

Another mechanism appearing in open-jet wind tunnels is the edge-tone feedback loop, first described by Rossiter [14] as some acoustic resonance in the investigated cavity. In open-jet wind tunnels this feedback loop is described as the shed vortices from the nozzle travel down the test section with about 65% of the reference velocity. As a vortex impinges on the collector a pressure shock appears and extends in the plenum. When this pressure jump hits on the nozzle edge it triggers a vortex to shed from the nozzle [7, 15]. The edge-tone frequency is therefore mainly depending on the test section length L_{TS} and calculated as:

$$f_e = \frac{1}{\frac{1}{m} \cdot \frac{L_{TS}}{0.65 \cdot U_{\infty}} + \frac{L_{TS}}{c - U_{\infty}}} \tag{6}$$

The mode number m and the speed of sound c are required for the calculation. Furthermore, the resonance frequencies of the wind tunnel plenum and the duct have to be taken into account. The volume frequency for a box-shaped volume is defined as

$$f_v = \frac{c}{2 \cdot \sqrt{\left(\frac{L_x}{n_x}\right)^2 + \left(\frac{L_y}{n_y}\right)^2 + \left(\frac{L_z}{n_z}\right)^2}} \tag{7}$$

with the lengths L and modes n in the three directions. The duct frequency is calculated with the length of the wind tunnel duct as

$$f_d = \frac{n \cdot c}{2 \cdot L} \tag{8}$$

The plenum Helmholtz resonance was verified by Beland [7]. In this model the air in the nozzle or collector can be seen as a mass and the air in the plenum as a spring. For the nozzle the frequency is calculated as:

$$f_H = \frac{r_{bn} \cdot c}{2 \cdot \pi} \sqrt{\frac{\pi}{V \cdot (L_{bn} + 0.5 \cdot \pi \cdot r_{bn})}} \tag{9}$$

2.3 Mode Decomposition

Proper Orthogonal Decomposition (POD) was used to catch the dominant modes of the flow fields. It is used in different fields of engineering and science to receive low-dimensional approximated results from complex higher-dimensional measurements. This approximation makes it possible to determine the dominant movements in flow systems.

The idea for POD was first published by Lumley [16] and further improved by Sirovich [17]. It is implemented in LaVision's PIV post-processing software, but can also be implemented in mathematic software like python or Matlab® easily. Good examples for the application of POD on fluid flows can be found in Meyer et al. [18] and Oberleithner et al. [19].

The velocity components u_{ij}, v_{ij} and w_{ij} of n two-dimensional planes with grid dimensions (i × j) are combined to one large matrix with (3 × i × j) rows and n columns:

$$\boldsymbol{U} = \begin{pmatrix} \boldsymbol{u}^1 & \cdots & \boldsymbol{u}^n \\ \boldsymbol{v}^1 & \cdots & \boldsymbol{v}^n \\ \boldsymbol{w}^1 & \cdots & \boldsymbol{w}^n \end{pmatrix} = \begin{pmatrix} u_{11}^1 & & u_{11}^n \\ \vdots & \cdots & \vdots \\ u_{ij}^1 & & u_{ij}^n \\ v_{11}^1 & & u_{11}^n \\ \vdots & \cdots & \vdots \\ v_{ij}^1 & & u_{ij}^n \\ w_{11}^1 & & u_{11}^n \\ \vdots & \cdots & \vdots \\ w_{ij}^1 & & u_{ij}^n \end{pmatrix} \tag{10}$$

Herein i = 1,. . ., P is the x-index and j = 1,. . ., Q is the y-index of the plane and n = 1,. . ., R is the number of snapshots in temporal order.

Subsequently this matrix **U** is multiplied with its transposed matrix to the auto-covariance matrix

$$\boldsymbol{C} = \boldsymbol{U}^T\boldsymbol{U} \tag{11}$$

After solving the eigenvalue problem

$$\boldsymbol{C}\boldsymbol{A}^n = \lambda^n\boldsymbol{A}^n \tag{12}$$

the R eigenvalues are ordered according to their size, with λ_1 being the largest and λ_S being the smallest eigenvalue. This guarantees that the first modes are the most important ones and show the large flow structures. The modes Φ^n for the POD are then reconstructed via

$$\boldsymbol{\Phi}^n = \frac{\sum_{k=1}^{R} \mathbf{A}_k^n \mathbf{u}^k}{\left\| \sum_{k=1}^{R} \mathbf{A}_k^n \mathbf{u}^k \right\|} \tag{13}$$

The POD coefficients combine the POD modes with the velocity field:

$$\mathbf{a}^k = \begin{pmatrix} \Phi^1 \\ \vdots \\ \Phi^S \end{pmatrix} \mathbf{u}^k \tag{14}$$

Finally the measured velocity fields can be reconstructed with a chosen number of modes. This leads to the elimination of smaller or less important flow-structures depending on the threshold mode. Besides that an evaluation of the dominant modes can help to understand the coherent flow structures.

2.4 Measurement and Simulation Methods

The investigation was carried out in the 20/25% model scale wind tunnel of the University of Stuttgart. It is a Göttingen type wind tunnel with a ¾ open jet and a test section length of 2.585 m. The nozzle has a height of 1.05 m and a width of 1.575 m, what leads to a hydraulic diameter of 1.26 m. Besides that, the tunnel is equipped with a 5-belt rolling road system with boundary layer suction and tangential blowing. More details to the wind tunnel and its latest upgrade in 2016 can be found in Wittmeier [20]. In this investigation the empty wind tunnel was operated without road simulation or boundary layer suction. The wind tunnel reference velocities U_∞ for the analysis were adjusted with the plenum method.

2.4.1 PIV

The used Stereo-PIV system from LaVision GmbH was set up to measure the shear layer in the wind tunnel centerline in various positions and window sizes. It consists of two Imager sCMOS CCD cameras, each with a resolution of 2560 × 2160 pixels and a dynamic range of 16 bit. The laser used to light up the 1 mm thick measurement slice is a 10 Hz fixed-frequency Nd:YAG laser from Quantel. It has a maximum energy of 400 mJ at the wavelength of 532 nm. The timing of the cameras and laser is controlled by a Programmable Timing Unit (PTU) and is depending on the velocity of the investigation and therefore the particle movement.

The laser and the camera system were installed on the negative y-axis side of the wind tunnel. The laser shot was steered from its initial direction in the positive y-axis to the xz-plane by an optical mirror and afterwards expanded to a slice via light sheet optic. The whole PIV system was mounted on the wind tunnel traversing system. This offered the possibility to move the system in the x-axis direction and measure various planes without recalibrating. The areas measured in the investigation are displayed in Fig. 1.

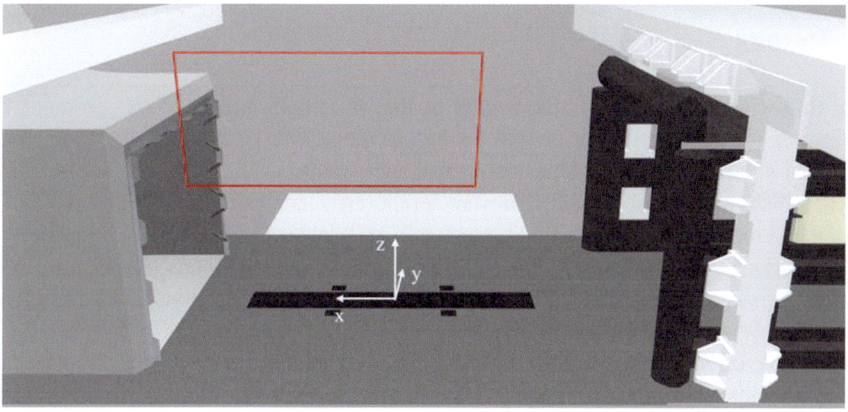

Fig. 1. The test section of the model scale wind tunnel of the University of Stuttgart with the areas of the shear layer investigated with PIV

After acquisition, the vector fields were calculated with the software DaVis 8.3.0 from LaVision. In a first step, background subtraction was used to damp out the background. The vector calculation was subsequently done on the workstation GPU via stereo cross-correlation in multi-pass with decreasing window size from 96 × 96 to 24 × 24 pixels and 50% overlap. The large planes for the POD and flow field analysis have a dimension of 655 × 570 mm with a resolution of 3.1 mm. The smaller planes for the shear layer development have a size of 290 × 245 mm and a resolution of 1.3 mm. For further detailed description of PIV, see for example [21]. After vector calculation, typical post-processing methods for PIV were applied. An allowable vector range fitting the particular velocity followed by the universal outlier detection as described in Westerweel et al. [22] were executed. Afterwards small gaps were closed by interpolation and 3 × 3 smoothing damps steep velocity gradients. To reduce small turbulences low-pass LES decomposition was applied [23].

2.4.2 Multi-hole Cobra Probe Measurements

The multi-hole Cobra Probe acquires the oncoming flow with an output rate of up to 5 kHz in an acceptance cone with an angle of 45°. The velocity is measured with an accuracy of ±0.3 m/s and a flow angle of ±1.0°. The probe was mounted to the traversing system and thus could be positioned in the largest part of the test section. The used output rate of the probe was 1.25 kHz.

The multi-hole probe measurements were processed in the center plane of the test section (y = 0) in different heights and x positions. The x position was varied from 1000 mm to –1100 mm in steps of 100 mm.

For the analysis of the measured data several methods were employed. The spectra were calculated with the Welch's power spectral density estimate in Matlab®. A Hamming window with a size of 4096 frames and 50% overlap was applied. The frequency resolution was therefore 0.30 Hz.

3 Results

The following section shows the results of the investigation. As a first comment, the definition for the used notation has to be clarified. The time-averaged results from Sect. 3.1 show the average values for a measured quantity. The time-resolved results in Sect. 3.2 show a transient and therefore temporarily analysis of the same results. In vehicle aerodynamics the terms "transient" or "unsteady" are often connected to investigations with unsteady side wind that arises on the road or is created in a wind tunnel. This is not the case in this study. There is no intended generation of turbulence in the wind tunnel by any means.

On the other hand, "time-resolved" PIV is normally connected to high-speed measurements with rates of several kHz. This is likewise not the case in this investigation. The term "time-resolved" is used here for the temporal analysis of the measured data.

3.1 Time-Averaged Results

The investigation focuses on the nozzle of an automotive wind tunnel. Therefore, the shear layer developing on the top of the nozzle is explored. The model scale wind tunnel of University of Stuttgart has no aero-acoustic usage, thus delta wings as displayed in Fig. 2(right) are mounted on the nozzle edges for vehicle force measurements. They are known to produce a certain high frequency noise, but reduce the low frequency buffeting of the plenum. Running the tunnel in day-to-day usage without delta wings would eventually result in serious damage to the structural integrity of the wind tunnel building. The performed measurements were executed with plain nozzle and with delta wings exceptionally for research reasons. This helps to understand the impact of turbulence generators such as delta wings in the shear layer of automotive wind tunnels.

The investigation is performed with different velocities between 30 and 60 m/s, correlating to Reynolds numbers based on the nozzle hydraulic diameter Re_{Dh} of 2.5 to 5.0 Million.

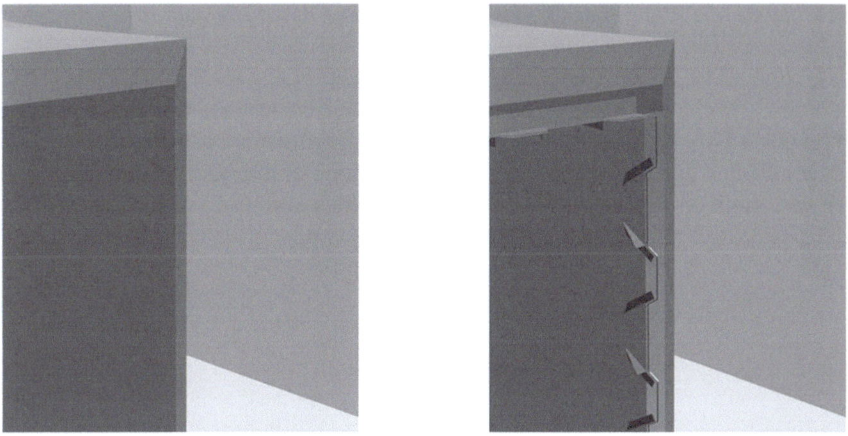

Fig. 2. Plain nozzle (left) and delta wings (right) at the nozzle end in the model scale wind tunnel of University of Stuttgart

The pressure distribution in the empty tunnel (plain nozzle) in a height of 150 mm is shown in Fig. 3. The pressure was acquired with the multi-hole cobra probe. All velocities indicate a raising pressure in the test section and therefore a decreasing velocity. This is what was expected a priori.

The behavior of the pressure is not linear. The measurement reveals the smallest raise in pressure over the test section for a velocity of 50 m/s. Especially the low velocity (30 m/s) shows a steep gradient with a buckle at x = 700 mm that results from the gap at the end of the center belt. This buckle is also visible in the 60 m/s case at x = 800 mm, but not in the 50 m/s measurement. In addition the nearly parallel

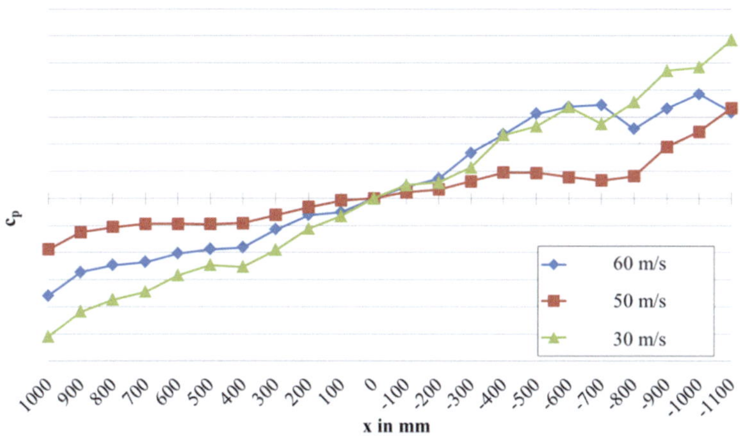

Fig. 3. Pressure coefficient distribution in the test section at z = 150 mm for different velocities for plain nozzle

development of the pressure for the 30 and 60 m/s cases between x = 100 mm and x = –600 mm can be noticed.

In contrast to that, the configurations with delta wings show a different pressure distribution. The deflection of the flow at the nozzle edge leads to an acceleration in the test section which is visible in Fig. 4 as a decreasing pressure coefficient. The pressure does not change between x = –100 and x = –800 mm until it finally rises reaching the collector. A small bend is displayed for all velocities at the end of the center belt.

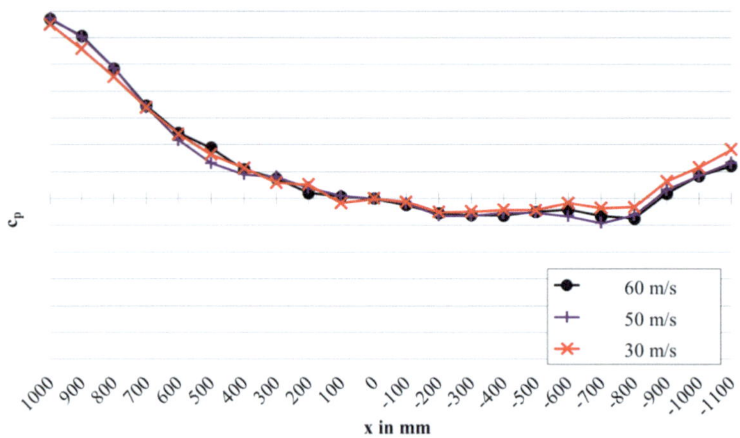

Fig. 4. Pressure coefficient distribution in the test section at z = 150 mm for different velocities with delta wings

Compared to the measurement with plain nozzle the pressure coefficients differ only in a small range. Both Figs. 3 and 4 show the same scale.

Another comparison between the plain nozzle and delta wings is displayed in Fig. 5. The turbulence intensity is plotted for all previous cases.

Obviously, the delta wings lead to a lower turbulence in the test section for all velocities. Another interesting observation is the oppositional behavior. For the measurement with delta wings, the highest velocity shows the lowest turbulence intensity, whereas Fig. 5 indicates large turbulence intensities for the plain nozzle especially in the rear part of the test section.

Just as determined for the even pressure coefficient distribution in Fig. 3 the 50 m/s case with the plain nozzle reveals small turbulence intensity compared to 30 and 60 m/s.

The x-velocity distribution and the vorticity thickness were determined from the PIV data. The reflections of the nozzle and the delta wings prohibited a reliable measurement directly at the nozzle edge, but a distance of 5% of a hydraulic diameter of the nozzle (63 mm) could be realized. The results for 63 mm and 630 mm in Fig. 6 show significantly larger vorticity heights for the plain nozzle near the nozzle exit. This means a steeper gradient for the case with delta wings.

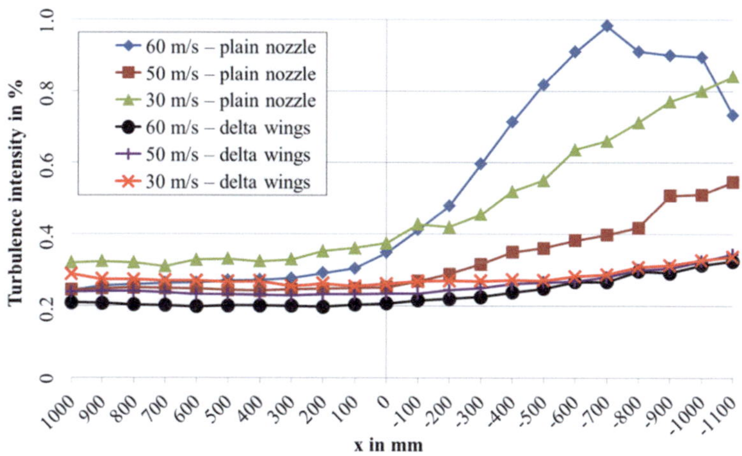

Fig. 5. Turbulence intensity in the test section at z = 150 mm for different velocities (plain nozzle and with delta wings)

A reverse trend can be recognized for the larger distance away from the nozzle. 630 mm behind the nozzle exit plane, the measurements with delta wings provide larger vorticity heights. This becomes more obvious in the x-velocity profiles.

In addition, it is clearly viewable in the right graph, that the delta wings force the shear layer further outwards than in the undisturbed case. The acceleration that was realized in the graph for the pressure coefficient suggests the opposite, a contraction of the test section and therefore an increase of the shear layer thickness.

Figure 7, from a CFD-Simulation which will not be discussed in detail, gives more insight into the flow topology. Iso-surfaces for a dimensionless absolute x-vorticity of 5.25 show the wakes of the two central delta wings. An x-normal surface 63 mm

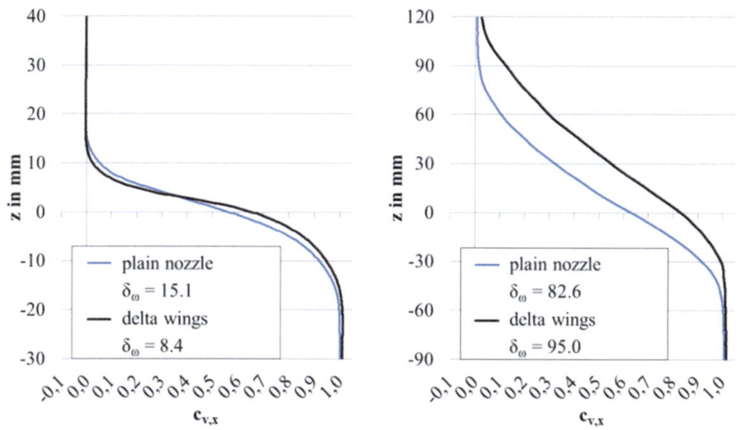

Fig. 6. X-velocity distribution in the shear layer 63 mm (left) and 630 mm (right) behind the nozzle exit plane at 60 m/s (plain nozzle and with delta wings)

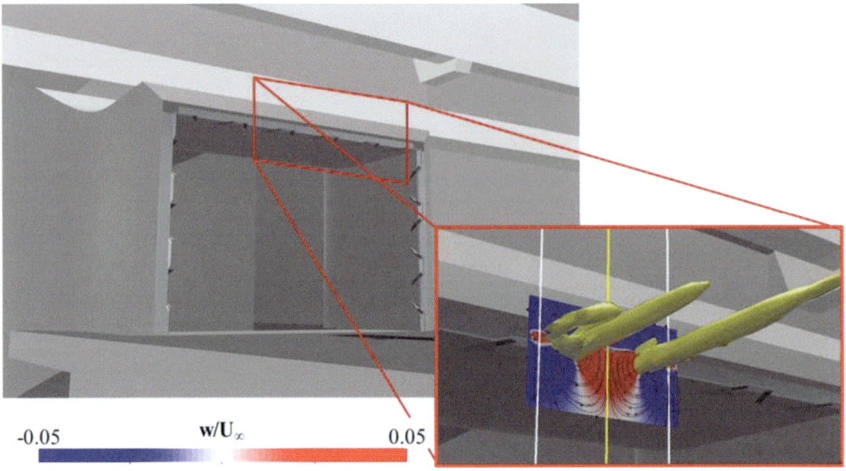

Fig. 7. Z-velocity and Iso-surfaces (x-vorticity = 5.25) in the wake of the central delta wings

behind the nozzle exit plane, colored with the z-velocity component, show a large upwards component of the flow in the center plane (yellow line, y = 0). Sideways the flow is pushed downwards between two converging delta wings (white lines).

3.2 Time-Resolved Results

The following sections show the time-resolved results of the measurements. Starting with the multi-hole probe spectra and continuing with the PIV data and its POD analysis.

3.2.1 Cobra Investigations

Figure 8 shows the measured spectra of the x-component of the velocity at the point (–500, 0, 750) mm for the three velocities 30 m/s (top), 50 m/s (middle) and 60 m/s (bottom). For a good comparison of the impact of the delta wings the left column illustrates the measurements with the plain nozzle. In contrast the right column displays the same velocities with delta wings.

This single geometric point shown in the graph is representative for the whole tests section with changing intensities of the peaks. Many other points were investigated as well, but the presentation of their results is too extensive.

All graphs result in a more or less distinct peak at a Strouhal number of 0.46 calculated with the hydraulic diameter of the nozzle (black dotted line). This correlates well with the theoretic value of vortex shedding of $Sr_{Dh} = 0.48$. This peak is also visible in the spectra for the z-component and is less dominant in the cases with delta wings.

A fixed frequency peak is found at 6.2 Hz for all cases in the velocity x-component (red line). As this frequency appears independently from the velocity it is most likely the duct frequency. The volume frequencies are significantly larger and the plenum Helmholtz frequency is about four times smaller.

The other observable peaks do not appear at a fixed Strouhal number or frequency. A massive peak at Sr_{Dh} = 0.74 (17.7 Hz) for 30 m/s is detected in most cases with the plain nozzle (orange dash-dotted line). They are always present in the x-component and also in the z-component in the rear part of the test section in all heights. In the lateral component they are most dominant in the central height and the height of the nozzle edge.

Besides that, moving peaks can be observed for the shown Fig. 8. They have no fixed frequency or Strouhal number and are pointed with a green dashed line. For 30 m/s the peak is found at Sr_{Dh} = 0.63 (15.0 Hz), for 50 m/s at Sr_{Dh} = 0.59 (23.5 Hz) and for 60 m/s at Sr_{Dh} = 0.54 (25.4 Hz). Calculated with the momentum thickness that was measured 10 mm behind the nozzle from the PIV data for the cases without delta

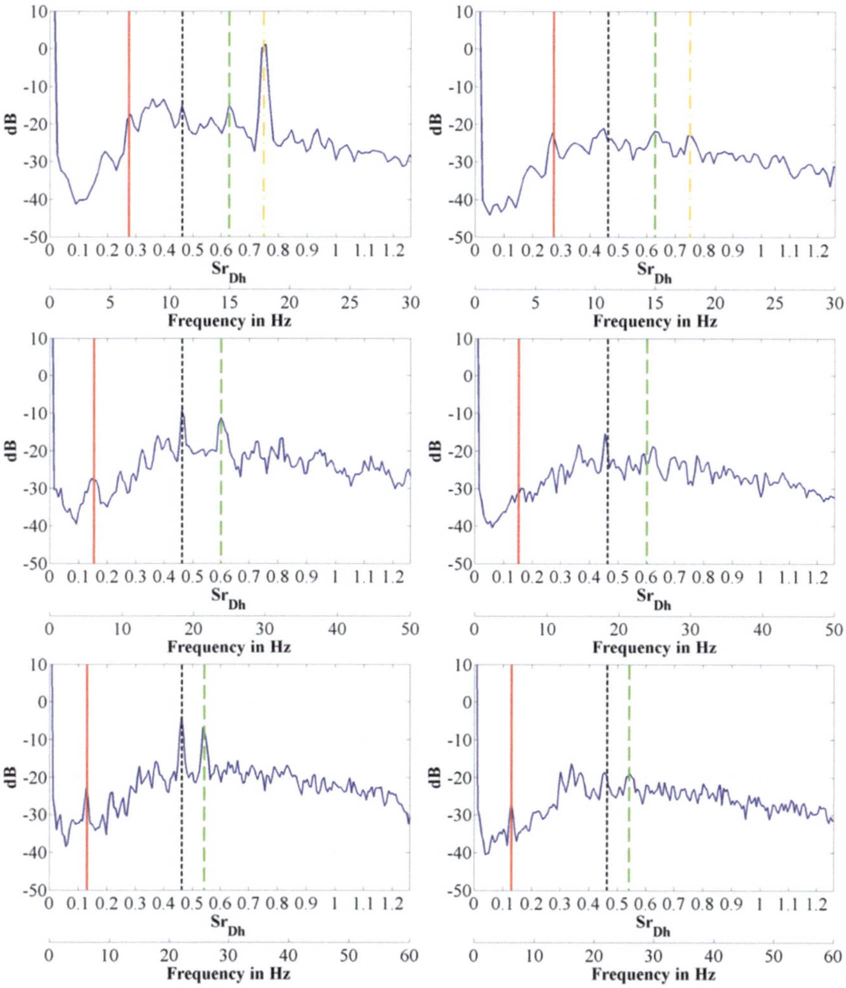

Fig. 8. Spectra of the velocity x-component at point (–500, 0, 750) mm for 30 m/s (top), 50 m/s (middle) and 60 m/s (bottom) (left column plain nozzle, right with delta wings)

wings a nearly constant Strouhal number of $Sr_{\delta 2} = 0.0011 - 0.0015$ was gained. This should be clarified in future investigations as there is an uncertainty in these PIV measurements because of background reflection. The boundary layer should be sized precisely with for example a total pressure probe.

Overall, the spectra for the cases with delta wings show less dominant peaks.

The pairing of vortices is a known event occurring in shear layers. As there is no halving of frequencies visible is the spectra, there is most likely no vortex pairing in the test section. This correlates with investigations in which vortex pairing was found in distances of over 1.75 hydraulic diameters behind the nozzle [24]. The model scale wind tunnel used in this research has a jet length of two hydraulic diameters. Therefore, vortex pairing may happen outside of the investigated planes or is prevented by the stagnation area in front of the collector.

3.2.2 PIV Investigations

On the height of the nozzle top edge, the changes in the x- and z-component of the velocity are a good indicator for passing vortices. Generally they are pictured as vortices separating from the nozzle edge and traveling in the shear layer with 2/3 of the reference velocity. To confirm this effect the following Fig. 9 shows instantaneous PIV vector fields for the 60 m/s measurement with the plain nozzle.

The first row displays the plane starting short behind the nozzle edge, the second and third row attach to the first and the second respectively. The planes have a height of around 400 mm and a width of around 650 mm. Thus, the measurements cover around 1.65 m in flow direction (1.3 nozzle hydraulic diameters).

Large vortices outside of the shear layer can be observed. In the following passages, these vortices are called outer vortices. On the height of the nozzle edge at $z = 1050$ mm waves are visible. The length of the waves can be found out by investigating classical vortex identification values as z-vorticity, swirl strength and lamb vector.

Generally, the distance between the peaks of these waves is growing with the distance from the nozzle. As a mean value the distance of the peaks develops from around $\lambda = 150$ mm in the front to around $\lambda = 550$ mm in the rear part of the investigated planes for a reference velocity of 60 m/s with the plain nozzle. Perhaps even larger distances appeared, but the size of the planes limited this investigation.

Together with the theoretic vortex travelling velocity from the edge-tone feedback of 40 m/s (65% of 60 m/s) frequencies between 75 Hz and 270 Hz are calculated. This means, that the shear layer vortices are most likely not responsible for the peaks investigated in the cobra measurements around 6 Hz to 26 Hz and for the classical vortex shedding frequencies.

As there were no two outer vortices identified in a single snapshot the distance between them has to be larger than the measured fields (650 mm). Together with the smaller travelling velocity outside of the shear layer, lower frequencies are a result for the outer vortices.

A comparison with the case with delta wings at the same velocity is analyzed in Fig. 10. They look similar to the previous case, but with a certain deviation to the upper part of the plane.

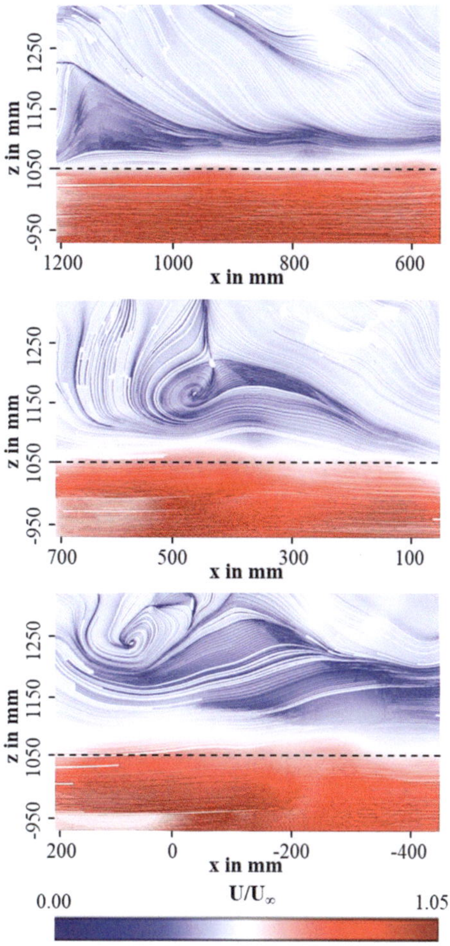

Fig. 9. Instantaneous velocity fields with streamlines for 60 m/s reference velocity at different x positions (plain nozzle, black dotted line is height of nozzle edge)

This is in good correlation to the time-averaged results, where a shift of the shear layer by the delta wings in the center plane was clearly visible. Again, large outer vortices can be detected outside the shear layer.

The position of the outer vortices was tracked in 100 instantaneous PIV streamline planes in all positions for the 60 m/s cases with plain nozzle and with delta wings. The results are plotted in Fig. 11. There are many ways to identify vortices [25] with different properties, but in this investigation, the focus is on the outer vortices. They were identified as closed streamlines of the velocity magnitude vectors in snapshots.

Vortices were found in around 20% of the snapshots. A constant drift away from the shear layer can be observed clearly. The earliest vortices can be identified at a

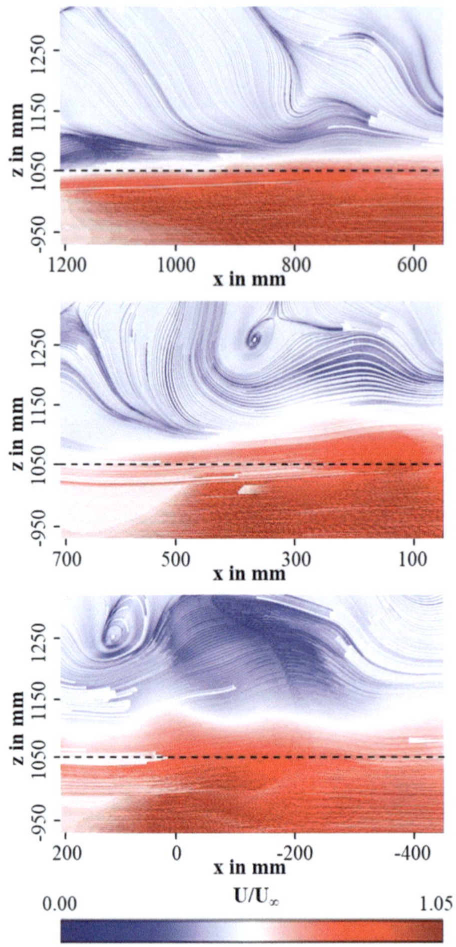

Fig. 10. Instantaneous velocity fields with streamlines for 60 m/s reference velocity at different x positions (with delta wings, black dotted line is height of nozzle edge)

position of around a half hydraulic diameter behind the nozzle exit plane around x = 650 mm.

The trend lines have similar inclinations of around 10% what correlates with the inclination of the outer boundary of the shear layer ($U/U_\infty = 0.01$). However, the trend line of the case with delta wings is shifted about 70 mm to the outside. The earliest vortices are again seen half a hydraulic diameter behind the nozzle.

The stagnation point of the upper part of the collector in the model scale wind tunnel is located at x = –1360 mm and z = 1220 mm. Therefore, the outer vortices do not hit the collector directly. They flow over the collector top part and imply a change in the x-velocity on the collector.

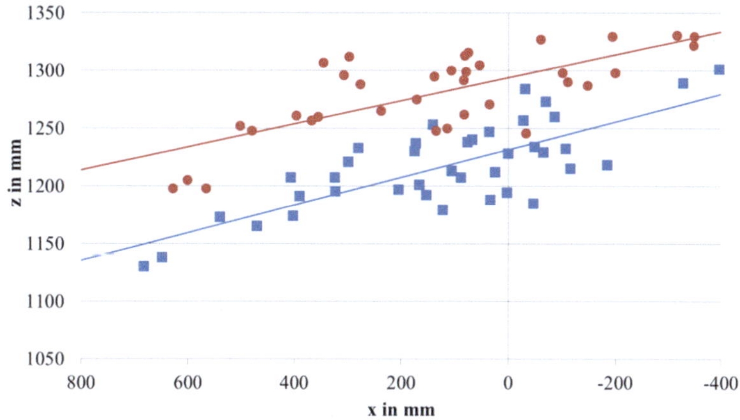

Fig. 11. Position of detected vortices outside of the shear layer and trend line for 60 m/s reference velocity (blue line plain nozzle, red line with delta wings)

3.2.3 POD Analysis

The POD analysis of the investigated cases results in vector fields and in the energies of the several modes. As remarked in Sect. 2 the modes are sorted by their values, hence the most energetic modes are the first ones.

For the different velocities and cases with plain nozzle and with delta wings the POD energies of the first four modes are displayed in Fig. 12. The plane investigated is the rear plane of the PIV investigation (x = 200 mm to x = –450 mm).

A significant difference is the deviation of the energies between cases with and without delta wings for decreasing velocity. The most dominant modes are the first and second mode of the 30 m/s case with plain nozzle. These modes include around 18 and

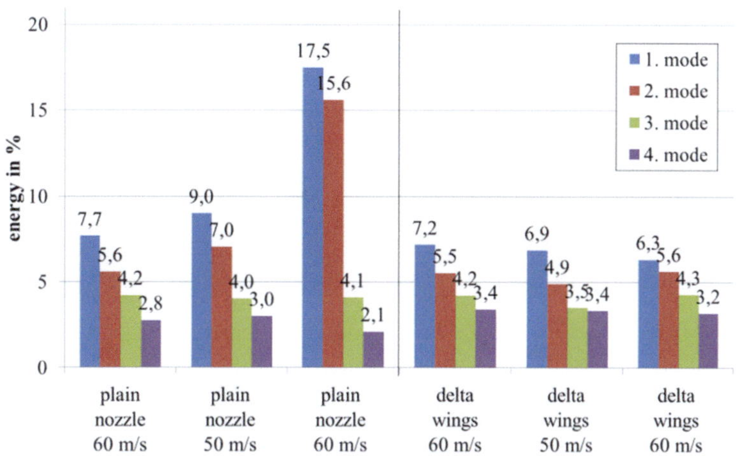

Fig. 12. Energy content of the first four modes for the rear plane from PIV

16% of the energy. In all cases with delta wings, the first two modes cover just 5 to 7% of the kinetic energy.

In Fig. 13 more insight into the structure of the first two modes is gained. Therefore, the vector fields of modes one and two are compared for the plain nozzle and velocities of 30 and 60 m/s.

Doubling the velocity leads to different results for the shown cases. On the left side two counter-rotating vortices are the only coherent structures to discover. They are positioned in the outer part of the shear layer and lead to large velocities in positive and negative z-direction. Compared to the snapshots the vortices are positioned nearer to the shear layer.

On the right side for the 60 m/s case there are also two vortices apparent. Again, they rotate in different directions but one of them is located in the outer part of the shear layer and the other one in the inner area. Thus, they imply a large velocity to the upper right or to the lower left respectively. The overall velocity level is lower in the 60 m/s case.

As in other POD investigations, two consecutive modes display a time delay of one fourth of a period time. There is no information about that period length, but the two counter-rotating vortices visible in Fig. 13 obviously show the same structures in different time frames.

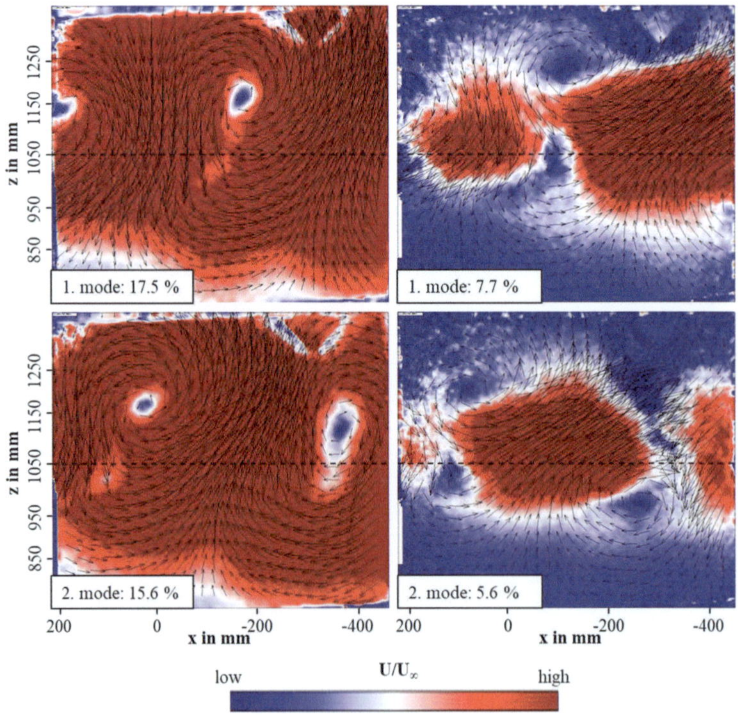

Fig. 13. Vector fields for the first two POD modes for cases with plain nozzle (left: 30 m/s; right: 60 m/s; black dotted line is height of nozzle edge)

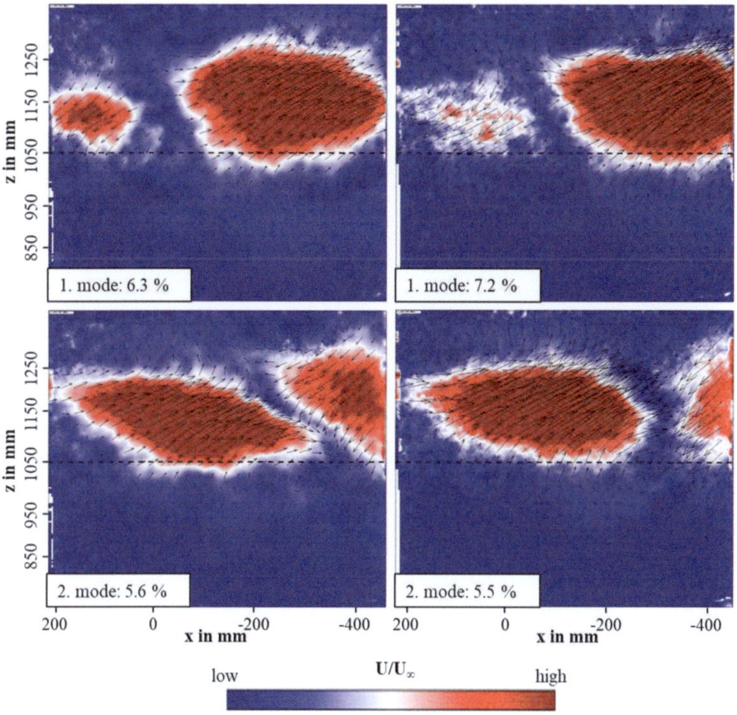

Fig. 14. Vector fields for the first two POD modes for cases with delta wings (left: 30 m/s; right: 60 m/s; black dotted line is height of nozzle edge)

The results with delta wings are plotted in Fig. 14. Some similarities with the 60 m/s case without delta wings can be observed. The flow in the center of the shear layer pointing to the upper right is implied by two counter-rotating vortex structures inside and outside of the shear layer. In contrast to the case with plain nozzle (60 m/s) the vortices are less distinct and the elongation in z-direction is smaller.

The most important finding is the large similarity between the different velocities. While the structures are very different for the plain nozzle with different velocities, the POD produces nearly the same vector fields in the measurements with delta wings.

The positions of the outer vortices from POD are shown in Fig. 15. Analog to previous figures the vortices are moving away from the shear layer in z-direction. The case with delta wings shows a steeper trend line and both trend lines are shifted about 40 mm towards the shear layer compared to Fig. 11 from snapshots.

Looking at the further modes reveals a certain coherent y-component (plane normal component) of the shear layer. This was not expected as existing theory focuses on the rotation around the y axis and often treats the shear layer as quasi two-dimensional.

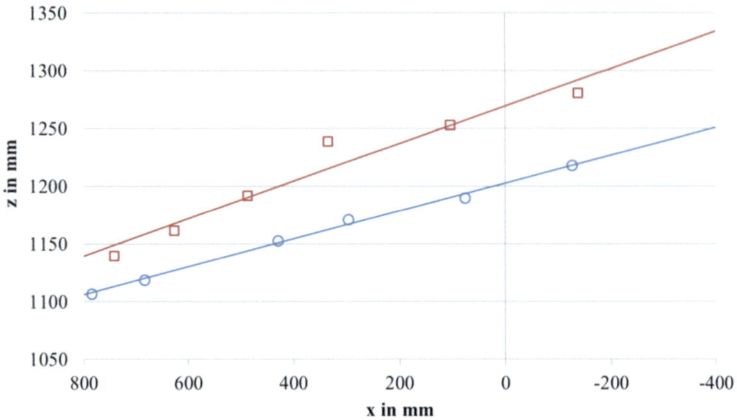

Fig. 15. Position of vortices and trend line from POD for the 60 m/s cases (blue line plain nozzle, red line with delta wings)

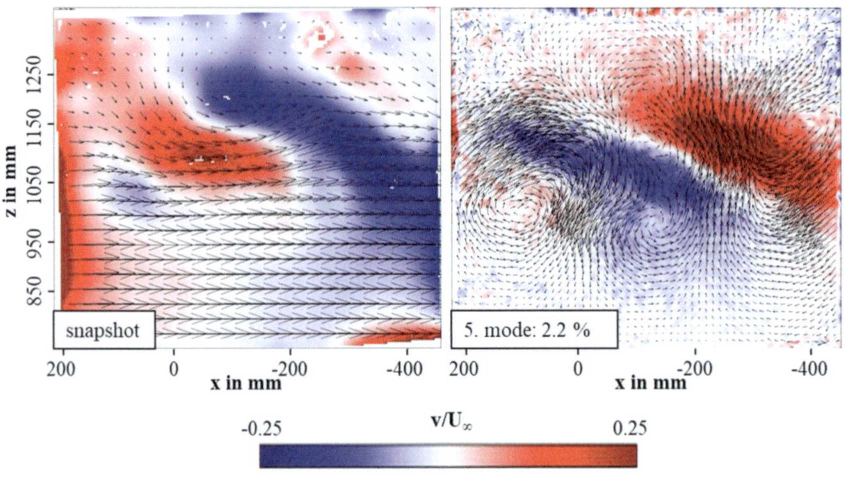

Fig. 16. Vector field with velocity y-component for snapshot (left) and fifth POD mode (right) for the 60 m/s case with plain nozzle

In Fig. 16 the velocity y-component of a snapshot and the fifth mode of the POD analysis are displayed for the case with delta wings at 60 m/s. Two slim areas with positive or negative y-velocity and a change of direction between the areas are visible. They extend from the upper left to the lower right of the plane.

All six investigated cases independent of velocity and delta wing configuration show this similar coherence in the y-component of the velocity. In the POD investigation, the coherence is revealed in all cases starting at mode five or six with an energy content of around 2% per mode.

This plane-normal velocity should be further investigated with an adapted PIV setup. Spanning up a xy-plane on the height of the upper nozzle edge will reveal more details of the flow structure.

4 Conclusions and Outlook

The shear layer flow in the model scale wind tunnel of University of Stuttgart was investigated at different velocities with the plain nozzle and with delta wings installed at the nozzle edges. A multi-hole probe and PIV were used to observe discrete points and two-dimensional fields in the center line of the wind tunnel.

Without delta wings, the time-averaged results show unexpected high deviations of the longitudinal pressure distribution for different velocities in the test section. This was not determined in the standard wind tunnel setup with delta wings installed. Therefore, the large turbulent structures in the shear layer, which are broken up by the delta wings, are responsible for the different pressure distributions.

The time-resolved measurements revealed several coherent effects. At a frequency of 6.2 Hz both nozzle configurations showed a peak in all investigated velocities. This peak seems to be connected to the wind tunnel duct. The Strouhal number $Sr_{Dh} = 0.46$ based on the nozzle hydraulic diameter, that is linked to vortex shedding, was verified in all configurations. Other peaks showed some connection to the velocity, but could not be assigned to a geometric entity.

The conception of a vortex ring separating at the nozzle edge and moving in the center of the shear layer was not confirmed. Instead, the PIV measurements show short waves arising from the nozzle edge. The wave length and amplitude grows with increasing distance from the nozzle. Those waves release large vortices on the outer part of the shear layer. The first vortices were detected at approximately half a nozzle hydraulic diameter behind the nozzle exit plane. After their formation, they move parallel to the outer boundary of the shear layer with an inclination of around 10%.

The POD analysis reveals differing modes for different configurations. Very intense vortices on the outer part of the shear layer were found for a velocity of 30 m/s without delta wings. Conversely, when the nozzle is equipped with delta wings hardly any dependency on the velocity and smaller vortices in the outer part of the shear layer could be demonstrated.

The PIV and POD investigations indicate a significant lateral component of the velocity. Because of the symmetry of the nozzle and test section, a quasi-two-dimensional flow property was expected.

Further research should investigate the axial expansion and appearance of the large vortices as well as the lateral pattern in the shear layer. This would also allow exploring the asymmetry of the flow forced by the delta wings. In addition, the investigated area should be extended further downstream to investigate the interaction with the collector. This could help to increase the understanding of the flow topology in the open jet of wind tunnels.

Acknowledgements. The authors like to thank the team of the model scale wind tunnel for the support and the colleagues from the vehicle aerodynamics and thermal management department for the assisting discussions.

Abbreviations

CFD Computational Fluid Dynamics
FKFS Research Institute of Automotive Engineering and Vehicle Engines Stuttgart
PIV Particle Image Velocimetry
POD Proper Orthogonal Decomposition

References

1. Mercker, E., Wiedemann, J.: On the correlation of interference effects in open jet wind tunnels. SAE Technical Paper 960671 (1996)
2. Mercker, E., Cooper, K., Fischer, O., Wiedemann, J.: The influence of a horizontal pressure distribution on aerodynamic drag in open and closed wind tunnels. SAE Technical Paper 2005-01-0867 (2005)
3. Mercker, E., Cooper, K.: A two-measurement correction for the effects of a pressure gradient on automotive, open-jet wind tunnel measurements. SAE Technical Paper 2006-01-0568 (2006)
4. Wickern, G., Schwartekopp, B.: Correction of nozzle gradient effects in open jet wind tunnels. SAE Technical Paper 2004-01-0669 (2004)
5. Hennig, A., Widdecke, N., Wiedemann, J.: A Study on Collector Interference Effects in Open Jet Wind Tunnels. Haus der Technik, Munich (2014)
6. Wickern, G., Heesen, W., Wallmann, S.: Wind tunnel pulsations and their active suppression. SAE Technical Paper 2000-01-0869 (2000)
7. Beland, O.: Buffeting suppression technologies for automotive wind tunnels tested on a scale model. In: Wiedemann, J. (ed.) Proceedings of the 6th FKFS-Conference on Progress in Vehicle Aerodynamics and Thermal Management, Stuttgart (2007)
8. Kuenstner, R., Potthoff, J., Essers, U.: The aero-acoustic wind tunnel of stuttgart university. SAE Technical Paper 950625 (1995)
9. Blumrich, R., Widdecke, N., Wiedemann, J., Michelbach, A., Wittmeier, F., Beland, O.: New FKFS technology at the full-scale aeroacoustic wind tunnel of University of Stuttgart. SAE Technical Paper 2015-01-1557 (2015)
10. Mercker, E.: On buoyancy and wake distortion in test sections of automotive wind tunnels. In: Wiedemann, J. (ed.) Proceedings of the 9th FKFS-Conference on Progress in Vehicle Aerodynamics and Thermal Management, Stuttgart, 2013
11. Schlichting, H., Gersten, K.: Boundary-Layer Theory. Springer, Heidelberg (2017)
12. Haigermoser, C., Vesely, L., Novara, M., Onorato, M.: A time-resolved particle image velocimetry investigation of a cavity flow with a thick incoming turbulent boundary layer. Phys. Fluids **20**, 105101 (2008)
13. Gong, J.: Grundlagenuntersuchung zur aktiven Beeinflussung der abgelösten Strömung. Dissertation Universität Stuttgart. Springer, Wiesbaden (2015)
14. Rossiter, J.E.: Wind-tunnel experiments on the flow over rectangular cavities at subsonic and transonic speeds. Ministry of Aviation, Aeronautical Research Council, Reports and Memoranda No. 3438 (1964)

15. Rennie, M., Kim, M.-S., Lee, J.-H., Kee, J.-D.: Suppression of open-jet pressure fluctuations in the Hyundai aeroacoustic wind tunnel. SAE Technical Paper 2004-01-0803 (2004)
16. Lumley, J.L.: The structure of inhomogeneous turbulent flow. In: Yaglom, A.M., Tatarski, V.I. (eds.) Atmospheric Turbulence and Radio Wave Propagation, Moscow (1967)
17. Sirovich, L.: Turbulence and the dynamics of coherent structures. Q. Appl. Math. **XLV**(3), 561–590 (1987)
18. Meyer, K.E., Pedersen, J.M., Özkan, O.: A turbulent jet in crossflow analysed with proper orthogonal decomposition. J. Fluid Mech. **583**, 199–227 (2007)
19. Oberleithner, K., Sieber, M., Nayeri, C.N., Paschereit, C.O., Petz, C., Hege, H.C., Noack, B. R., Wygnanski, I.: Three-dimensional coherent structures in a swirling jet undergoing vortex breakdown: stability analysis and empirical mode construction. J. Fluid Mech. **679**, 383–414 (2011)
20. Wittmeier, F.: The recent upgrade of the model scale wind tunnel of University of Stuttgart. SAE Technical Paper 2017-01-1527 (2017)
21. Raffel, M., Willert, C., Wereley, S., Kompenhans, J.: Particle Image Velocimetry. Springer, Heidelberg (2007)
22. Westerweel, J., Scarano, F.: Universal outlier detection for PIV data. Exp. Fluids **39**, 1096–1100 (2005)
23. Adrian, R.J., Christensen, K.T., Liu, Z.-C.: Analysis and interpretation of instantaneous turbulent velocity fields. Phys. Fluids **29**, 275–290 (2000)
24. Zaman, K.B.M.Q., Hussain, A.K.M.F.: Vortex pairing in a circular jet under controlled excitation. J. Fluid Mech. **101**, 449–491 (1980)
25. Jeong, J., Hussain, F.: On the identification of a vortex. J. Fluid Mech. **285**, 69–94 (1995)

New Vehicles

Mercedes-AMG GTR: Aerodynamics for the Record

Gustavo Estrada(✉)

Daimler AG, Daimlerstr. 1, 71563 Affalterbach, Germany
gustavo.estrada@daimler.com

Abstract. Mercedes-AMG continues to grow. The sports car and performance brand of Mercedes-Benz expanded the top end of its product range with the introduction of the new Mercedes-AMG GT R. Never before has Mercedes-AMG packed so much motorsport technology into a production vehicle than it has in the new AMG GT R. The challenge for the aerodynamic engineers was the development of a unique car with special requirements: distinctive proportions, clean coupe design, front-mid-twin-turbo V-8 engine rated at 430 kW/585 hp, singular package, light weight, transaxle, extreme cooling requirements and above all, the goal of being the fastest of its class on the world's most demanding racetrack. The result is an intelligent, distinctive and innovative combination of aerodynamic features that fits the complex requirements of this vehicle. The interaction of the active aerodynamic features provides the right aerodynamic performance for each driving situation. This allows the combination of driving dynamics of a Mercedes-AMG GT3 race car with the everyday practicality of the Mercedes-AMG GT, assuring race circuit performance and low fuel consumption. The result is a lap time of 7:11 min (Journal Sport Auto 1/2017) at the Nordschleife. Benchmark.

1 About the Mercedes-AMG GT

The AMG GT R is the new spearhead of the AMG model range. Featuring pioneering technologies, the AMG GT R was designed with racetrack use firmly in mind. Its development was largely based on vast motor-racing experience in AMG GT3 customer sport and the German Touring Car Masters. Figures such as 3.6 s from zero to 100 km/h and a top speed of 318 km/h, combined with the outstanding driving dynamics, will undoubtedly translate into extremely fast laps on the racetrack.

Sharpened racetrack performance comes courtesy of the intelligent lightweight construction incorporating aluminum and carbon, specific reinforcements in the body shell, the uprated twin-turbo V-8 engine with the specifically adapted seven-speed dual-clutch transmission, the innovative aerodynamics with active components, and the new suspension with active rear-wheel steering and uniball joints.

The concept with front mid-engine and transmission in transaxle configuration on the rear axle, a tried-and-tested feature of the AMG GT and AMG GT S, makes for a beneficial rear-based weight distribution of 47.3 to 52.7. In conjunction with the vehicle's low centre of gravity, this translates into extremely agile handling and permits

J. Wiedemann (ed.), *Progress in Vehicle Aerodynamics and Thermal Management*,
https://doi.org/10.1007/978-3-319-67822-1_8

high cornering speeds. With a power-to-weight ratio of 2.66 kg/hp, the Mercedes-AMG GT R takes up pole position in its segment.

The data at a glance:

	Mercedes-AMG GT R
Engine	4.0-litre V-8 with twin turbochargers and direct injection
Displacement	3982 cm^3
Output	430 kW (585 hp) at 6250 rpm
Peak torque	700 Nm at 1900-5500 rpm
Driven wheels	Rear-wheel drive
Transmission	AMG SPEEDSHIFT DCT 7 speed dual-clutch transmission
Fuel consumption - urban/extra-urban/combined	15.0/9.2/11.4 l/100 km
Combined CO_2 emissions	259 g/km
Efficiency class	G
Weight (DIN/EC)	1555*/1630** kg
Power-to-weight ratio	2.66*/2.79** kg/hp
Acceleration 0-100 km/h	3.6 s
Top speed	318 km/h

*Kerb weight according to DIN, not including driver and luggage; **Kerb weight according to EC, including driver (75 kg)

2 Aerodynamics

The aerodynamic development of a Mercedes-AMG GT is pushed by multiple challenges. First of all, the vehicle's proportions as shown in Fig. 1 mean a hard starting point to reach the lift and drag targets. From the thermal point of view, considering the package characteristics and the cooling requirements, it demands intelligent solutions in order to reach the hotspots with fresh air. Furthermore, the goal of the conception of light weight solutions pushed the aerodynamic development to its limit.

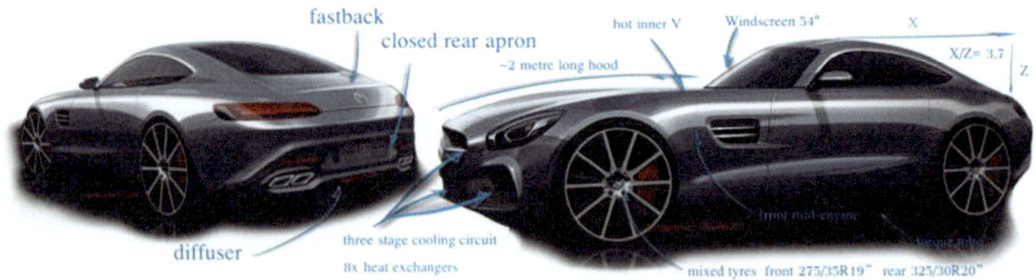

Fig. 1. Silhouette and proportions of a Mercedes-AMG GT

3 The Front End

Challenge: extreme cooling requirements, three stage cooling circuit, eight radiators, twin turbochargers inside the V configuration ("hot inner V") vs. low drag C_D *and reduced frontal lift* C_{LF}.

The low-slung front section and the forwards-inclined radiator grille create a distinctive "shark nose". This shape lowers the vehicle's stagnation-pressure point, enhancing the flow of cooling air and the car's aerodynamic performance. Two air inlets in the centre and one outer air inlet each side at the wheel arches are necessary to fulfil the cooling requirements. The vertical grille's bars at the nose of the car, trademarked as Panamericana Grille, are designed as tapered profiles to force air into the radiators, and there is nearly no air recirculation to the engine compartment (Fig. 2).

Fig. 2. Mercedes-AMG GT R front end

For optimum power output even when outside temperatures are high, Mercedes AMG uses indirect air-to-water intercooling. With optimum flow of air and water, the intercoolers have a separate, two-stage low-temperature water circuit. The first cooler stage involves two parallel coolers in the left and right wheel arch. The outer air intakes in the front fascia ensure that the increased cooling-air requirements of the AMG GT R drive system are met (Fig. 3).

Fig. 3. Cooling system (without main radiator) and engine of a Mercedes-AMG GT R

A second radiator stage includes a large radiator at the front of the vehicle. The downstream water-cooled intercoolers ensure that the charge air compressed and heated by the turbochargers is cooled effectively prior to entering the combustion chambers. It therefore remains at a constantly low level even under full load. A large radiator at the car's front end ensures controlled cooling of the water circulating in the low-temperature circuit.

To fulfil the requirements of frontal lift (C_{LF}) reduction while increasing of mass flow for the cooling system, the wide front splitter was integrated in the design. In this way, these elements have an inherent technical benefit and a proportion that fits the vehicle's design.

Additional Air Curtains on the outside of the front fascia guide the air specifically towards the wheel arches, optimizing the flow properties in this area. The result is an improved C_D value and a further reduction of frontal lift (C_{LF}).

3.1 All-New Active Aerodynamics Profile in the Underbody

Challenge: reduce frontal lift (C_{LF}) without increasing drag (C_D) and without modifying the vehicle's shape or proportions vs. light weight.

A special engineering feature is the completely new active aerodynamics profile, which is concealed almost invisibly in the underbody in front of the engine and thus blends in harmoniously with the overall silhouette.

Multiple functions are combined when the device is deployed: a reduction of frontal lift (C_{LF}); an increase of engine cooling and also an increase in brake cooling. This multi-purpose feature reduces the drag (C_D) at the same time without increasing of the rear lift (C_{LR}) of the car (Fig. 4).

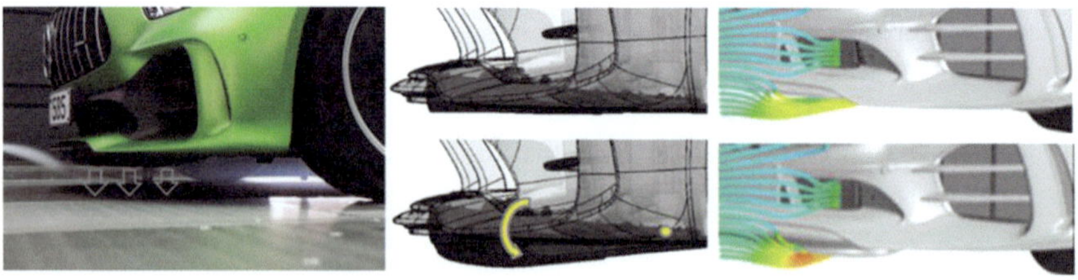

Fig. 4. Aerodynamic profile in the underbody

The aerodynamic profile deployed on the underbody at a speed of 80 km/h in "RACE" mode, and at 120 km/h in "Comfort", "Sport" and "Sport Plus" modes. This carbon fibre element, which only weighs around two kilograms, automatically moves approximately 40 mm down (Fig. 5).

This process involves considerable change in the airflow and results in what is known as the Venturi effect, which additionally presses the car onto the road and reduces the front-axle lift by around 40 kg at 250 km/h. When the AMG GT R in "RACE" mode drives slower than 60 km/h or slower than 80 km/h in "Comfort", "Sport" and "Sport Plus" modes, the aerodynamics profile moves back in. To protect against damage, the component is spring-mounted and can therefore flex upwards easily.

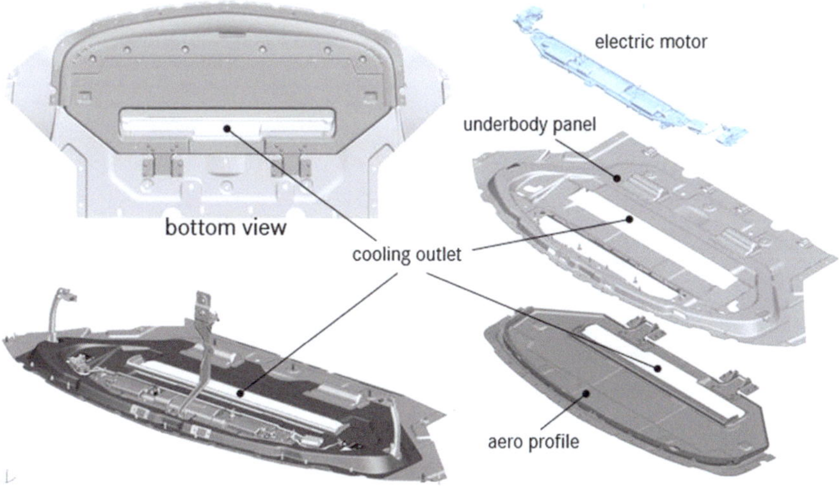

Fig. 5. Description of the underbody aero profile

The driver can feel the advantages of this aerodynamic feature at the steering wheel: the AMG GT R is even more precise to steer when cornering at high speed and exhibits even better directional stability. Specially during fast cornering and under high lateral acceleration, the AMG GT R delivers far more agile response with clear steering-wheel feedback while remaining easily controllable at all times.

The efficiency of this feature can be proved comparing it to standard features such as a splitter and flics (Fig. 6). In order to generate the same lift forces and lift balance, it would be necessary to mount a 60 mm (almost 24 in.) front splitter and flics, increasing the drag coefficient significantly.

Fig. 6. Splitter proportions and flics to generate similar frontal lift (C_{LF}) as the active aero system on the underbody

When the electrically operated profile is extended, the radiator air outlet opens at the front end and precisely guides the air flow towards the double rear diffuser, which therefore also benefits from an optimal flow of air (Fig. 7).

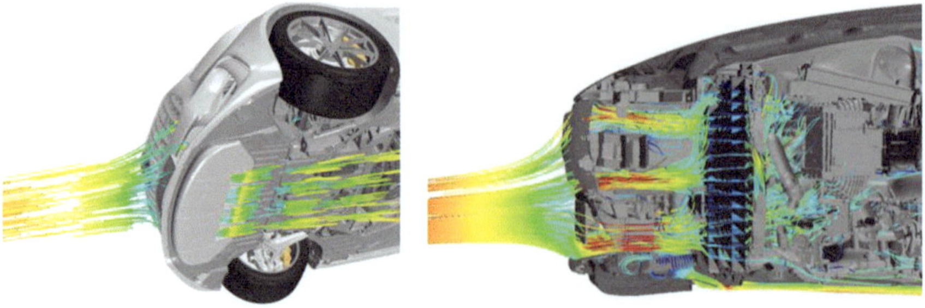

Fig. 7. Flow through the cooling system when the aerodynamic profile is deployed and front louvres are opened

This improves the handling stability of the rear axle while reducing the temperature level of the hotspots at the rear. At the same time, the aerodynamic package ensures optimized braking power by routing more cold air to the wheel discs specifically. When the profile is deployed, the braking-duct-diffusor on the underbody increases its depth by almost 50%, directing the necessary clean cold air to the brakes.

3.2 Louvres Control the Airflow: The Active Air Management System

Challenge: High engine power and high cooling needs vs. drag reduction in combination with frontal lift control. Cold air "just when you need it".

Another technical highlight improving the aerodynamics of the AMG GT R is the active air management system. This system has vertical louvres positioned in the lower area directly behind the front fascia. These louvres are electronically controlled and can be opened and closed in approximately one second by an electric motor to improve the airflow and consequently the aerodynamic performance (Fig. 8).

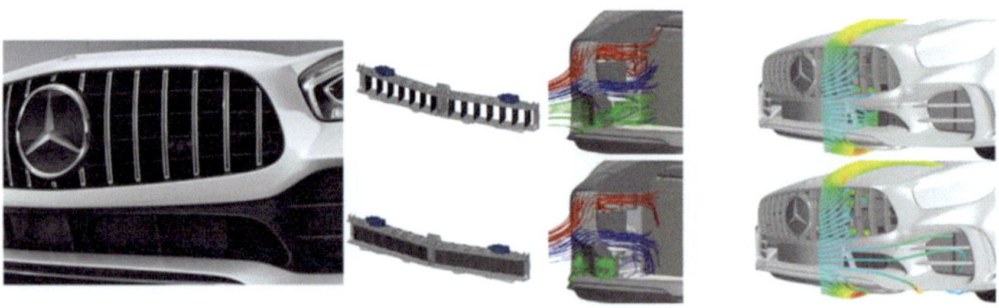

Fig. 8. Frontal louvres for thermal requirements and control of aerodynamic coefficients

The louvres are normally closed - including at top speed, during braking and when cornering at high speed. This position lowers drag and makes it possible to route the air to the underbody to reduce front lift even further. Only once predefined components have reached certain temperatures and the demand for cooling air is particularly high, the louvres quick open and allow the maximum amount of cooling air to flow to the heat exchangers. This calls for highly intelligent and fast control. The direction of rotation of the louvres is optimized for the fast response at the heat exchangers.

4 Emphasis on Width and Aerodynamic Optimization: The Rear End

Challenge: Coupe form, clean and compact shapes, a trunk, high temperatures produced from the exhaust system, closed underbody vs. no big openings or meshes to exhaust the heat at the rear apron and the requirement of 150 kg of downforce at Vmax.

4.1 Diffuser

The wider design of the rear fascia with large outer vent openings and vertical swaging improves the airflow at the rear end, as does the double diffuser. A further distinguishing feature of the AMG GT R is the large exhaust tip with centre spar, which is centrally positioned in the rear fascia. It is flanked by two more black exhaust tips on left and right in the diffuser (Fig. 9).

Fig. 9. Rear view and exhaust system

As in Fig. 10 described, the hot surroundings of the titanium muffler is exhausted with fresh and clean air coming from the underbody.

This flow was accelerated and directed thanks the aerodynamic profile at the front of the vehicle and it is redirected to the mufflers without any physical conducts but based on pressures differences.

The overall heat dissipation of the silencers is in this way exhausted by the double diffuser while driving.

But the function of the double diffuser wouldn't be possible without the interaction with the rear wing.

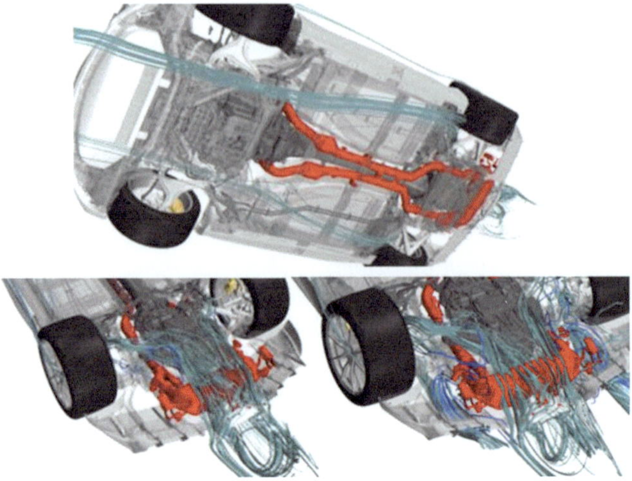

Fig. 10. Streamlines started at upper side of titanium muffler

4.2 The Rear Wing

The large rear wing is rigidly mounted on the hatch and increases the downforce on the rear axle. Depending on the type of use or racetrack conditions, the precise angle of the blade can be adjusted manually. This is possible because the active underbody aero system was developed to have almost no negative effect on the rear end's dynamics. This means that the wing doesn't need to compensate the reduction of the frontal lift C_{LF}. This advantage means a reduction of weight and complexity and the chance for the drivers to feel free to adjust the rear wing to their liking in a range of just 10° (Fig. 11).

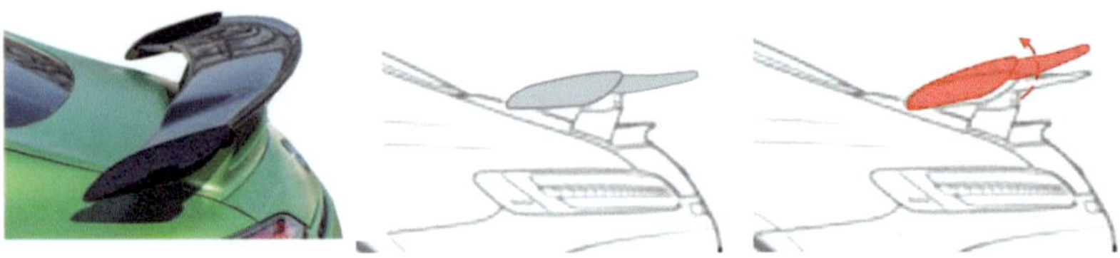

Fig. 11. Manually adjustable rear wing

5 The Result: Some Aerodynamic Values

The impact on drag and lift of the three aerodynamic active features is shown in Fig. 12. The results are given in reference to the vehicle in the following configuration:

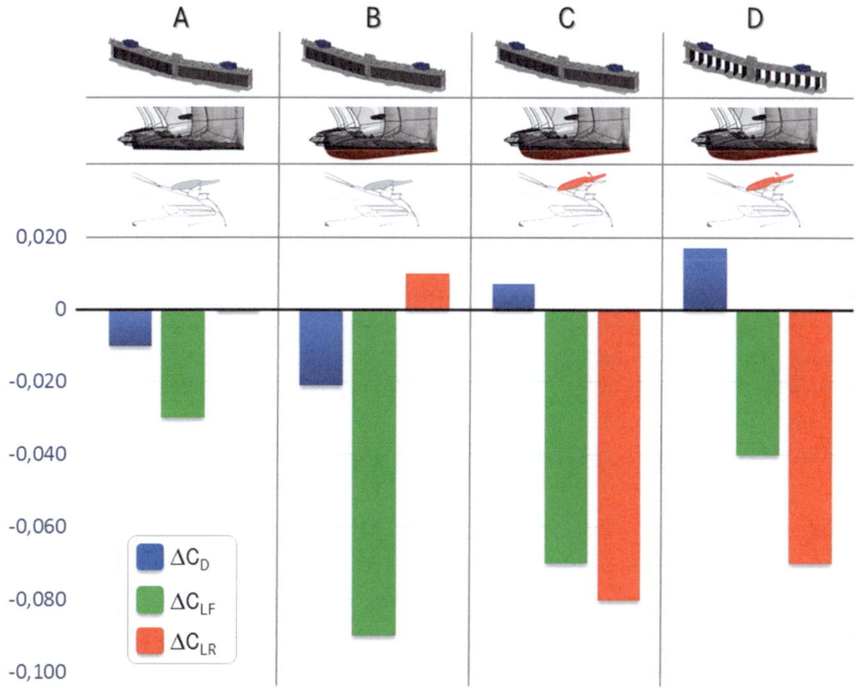

Fig. 12. Aerodynamic effects of the AMG GT R features referred to the configuration: louvres opened + underbody profile retracted + rear wing low

Reference: louvres opened + underbody profile retracted + rear wing low

The **Configuration A** represents the closing of the active air management system (louvres)

A: louvres closed + underbody profile retracted + rear wing low

showing the reduction of drag (C_D) and frontal lift (C_{LF}) but no increase of rear lift (C_{LR}). This configuration is possible to drive at low speed (lower than 80 km/h in "RACE" mode, or 120 km/h in "Comfort", "Sport" and "Sport Plus" modes) providing sufficient ground clearance and meeting daily practicality.

At the **Configuration B**,

B: louvres closed + underbody profile deployed + rear wing low

the profile at the underbody is deployed, producing a massive reduction of frontal lift (C_{LF}) and even a higher reduction in drag (C_D). The rear axle loses marginal some downforce, but without affecting the rear end's dynamics. At this point, the vehicle is ready to drive at higher lateral dynamics even with the wing in the lower position. The driver is able to modify the wing (adapting the lift balance) according to their needs.

At the **Configuration C**

C: louvres closed + underbody profile deployed + rear wing high

the rear wing is adjusted at its max angle. The rear downforce (C_{LR}) increases to its maximum. This configuration can be driven up to Vmax. The louvres will open in case more cooling air is needed (**Configuration D**) and they will close again when the required engine and hotspots temperatures are reached.

It is clearly to see the interaction of the active features, giving each driving situation the right aerodynamic performance. The intelligent, distinctive and innovative combination of aerodynamic devices fits the vehicle to reach its complex requirements. This allows the combination of driving dynamics of a Mercedes-AMG GT3 race car with the everyday practicality of the Mercedes-AMG GT, assuring race circuit performance and low fuel consumption. The new AMG GT R therefore offers maximum grip, perfect drivability and a powerfully proportioned design that integrates the aerodynamic functions harmoniously.

The Aerodynamics Development of the New Land Rover Discovery

Sébastien Chaligné(✉), Ross Turner, and Adrian Gaylard

Jaguar Land Rover, Banbury Road, Gaydon, Warwick CV35 0RR, UK
schalign@jaguarlandrover.com

Abstract. With ever stricter emissions regulations and the upcoming certification procedure defined in WLTP, the importance of aerodynamics for OEMs is increasing. This paper presents the aerodynamics development of the fifth generation Land Rover Discovery and gives an insight into the recent migration towards the use of new tools and processes to ensure future vehicles' compliance is achieved.

In the early development stages, numerical methods were exclusively used to optimise the main proportions of the vehicle, as well as to understand sensitivities and draw a road-map to attain the aerodynamics attribute targets. A full-scale test property ("Aerobuck") was then built and tested in the fixed-ground MIRA Full Scale Wind Tunnel (FSWT) to optimise specific areas in order to reduce the drag coefficient. These tests were done in combination with an extensive use of CFD to enable a better understanding of the flow fields and mechanisms involved. Finally, the development of an aerodynamically optimised wheel enabled the lowest drag coefficient to date for a Land Rover SUV, $C_D = 0.33$.

Although certified in the MIRA FSWT, the validation process has also seen prototypes tested in the moving-ground FKFS Aeroacoustic Wind Tunnel (AAWT) benefiting from 5-belt ground simulation. A direct comparison between experimental and numerical results has also been made; they generally show good agreement between the two tools except for the prediction of so-called "cooling drag".

1 Introduction

With the introduction of new homologation regimes and stricter requirements for CO_2 emissions developing a new full-size SUV presents aerodynamic challenges. The New Discovery is Land Rover's fifth generation of this premium SUV. It replaces the Discovery 4, which was developed from the closely related Discovery 3, a vehicle that debuted in 2004.

The outgoing vehicle (Fig. 1) illustrates the challenges. Designed to accommodate up to seven full-sized seats, with stadium seating and a command driving position, the exterior is characterised by largely vertical surfaces and a stepped roof line. The requirement for class-leading off road performance drove short front and rear overhangs. As a consequence the Discovery 3 was a bluff two-box SUV with a

The original version of this chapter was revised: Chapter title has been changed. The erratum to this chapter is available at https://doi.org/10.1007/978-3-319-67822-1_18

J. Wiedemann (ed.), *Progress in Vehicle Aerodynamics and Thermal Management*,
https://doi.org/10.1007/978-3-319-67822-1_9

projected frontal area (A) of 3.15 m^2 and a drag coefficient (C_D) on launch of 0.41 [1], which reduced to less than 0.40 over subsequent model years. It was also developed using Jaguar Land Rover's previous generation of technical approaches: a test-lead process focussed on fixed ground wind tunnel testing supported by a limited application of CFD [2, 3].

The development programme for the 2017 New Discovery aimed to maintain capability, such as off road performance and seating for seven adults, whilst delivering

Fig. 1. The Land Rover Discovery 3

a new Design language compatible with future CO_2 requirements and the homologation philosophy embodied in WLTP. However, this left little scope for frontal area reduction and meant that aerodynamic improvement had to be approached via improved shape efficiency, i.e. reduced C_D.

The aerodynamic development of the New Discovery illustrates both the transition to new regulatory approaches and development processes, with CFD leading the development process [4] and wind tunnel testing transitioning to the use of a moving-ground facility.

2 Aerodynamic Development Process

The development process of the New Discovery consisted of four main stages: strategy, delivery, launch and production phases. The aerodynamic process is essentially embedded within the first three stages which will be described in the following section.

Throughout the development a number of tools have been used to assess the aerodynamics of the vehicle. In the strategy phase, the CFD software EXA PowerFLOW was exclusively used to optimise the main proportions of the vehicle, as well as to understand sensitivities and draw a road-map to attain the aerodynamics attribute targets. This is a commercially available CFD code, which provides a Very Large Eddy Simulation (VLES) turbulence model based on a Lattice Boltzmann (LB) solver [5, 6]. This inherently unsteady approach resolves the flow and any large scale unsteadiness on a regular cubic lattice with nested, hierarchical regions of enhanced resolution. The effect of unresolved unsteady motions are accounted for by modifying the behaviour of the LB simulation, including an effective turbulent relaxation time, calculated via the RNG κ-ε transport equations [7]. It is a well-known development tool in the automotive industry and has been employed previously by Jaguar Land Rover in the aerodynamic development of vehicles such as the original Jaguar XF [8], Range Rover Evoque [3] and Jaguar XE [4]. The Discovery was also the first programme to benefit from a deployed soiling CFD simulation method [9]. Figure 2 shows that over half of the total number of CFD simulations was run in this early design phase.

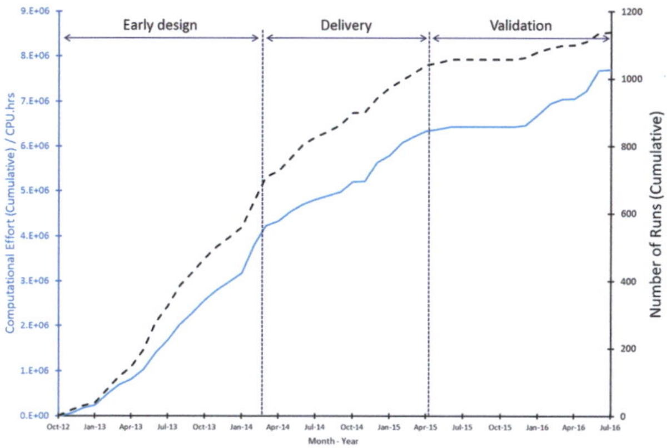

Fig. 2. Number of CPU hours used throughout the vehicle development

In the delivery phase, CFD was still used but in combination with experiments in the MIRA Full Scale Wind Tunnel (FSWT). This tunnel has a closed test-section open-return (NPL) lay-out with an informal return-path generated by its location in a large hanger. The test section floor is fixed, with no floor boundary layer control [10] and force measurements are made via an under floor balance, with the vehicle's non-rotating wheels sat on four pads reacting their load to the balance. The influence of blockage on force and pressure measurements is accounted for via a one-dimensional continuity correction [11]; an additional correction is made to the measured drag coefficient accounting for the longitudinal pressure gradient.

Validation prototype vehicles became available in the launch phase enabling the aerodynamic performance to be signed off against the target. Testing of these prototypes was undertaken primarily in the FSWT for consistency with development targets; however, some exploratory testing was also conducted in the AAWT to aid understanding in preparation for the upcoming changes in certification procedures. The Aeroacoustic Wind Tunnel of Stuttgart University has been extensively developed for automotive aerodynamic purposes over several decades [12]. The test vehicle is set on a 5-belt system with restraint struts holding it in position [13]. This set up enables a better representation of the on-road conditions compared to the FSWT by simulating the relative motion between the vehicle and road, as well as wheel rotation.

2.1 Early Design Characterisation

The early design phase consisted of two main parallel work-streams, both completed exclusively numerically. First the volume and proportions of the vehicle were optimised to ensure compatibility between the Design vehicle vision and aerodynamic attribute targets, whilst identifying the drag sensitive areas around the vehicle and develop a road-map designed to deliver the aerodynamic drag target.

Second, characterisation work enabled drag reduction opportunities to be identified and quantified. These ideas were progressed with the Engineering and Design teams to develop and refine the roadmap to target, as per the example shown in Fig. 3.

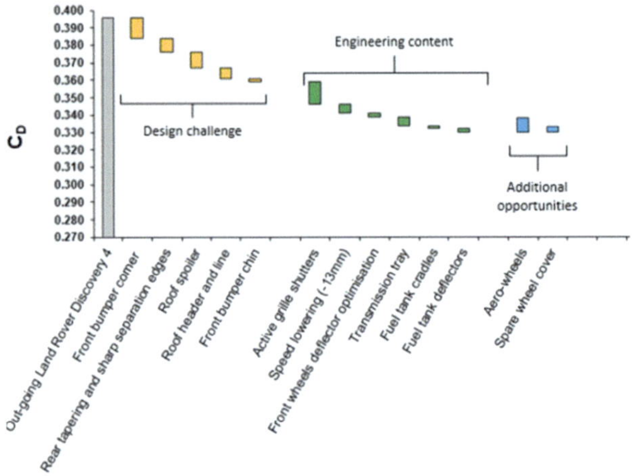

Fig. 3. Example of early design road-map to deliver the aerodynamic attribute drag target

It was clear from the start that to deliver the requisite step-change in drag a significant departure from the exterior Design of the out-going Discovery would be required. Collaboration with Design started very early in the programme, enabling the team to influence the initial exterior themes. For example, work was done to minimise the drag impact of the stepped roof – enabling the preservation of a signature Discovery feature that improves interior roominess. Other important areas of development were the front corners, rear fenders and tailgate spoiler.

2.2 Delivery Phase

Once a single Design Theme was chosen and the engineering content finalised, detailed optimisation began. CFD was used to understand geometric sensitivities and guide development. The full-size "aerobuck" test property, shown in Fig. 4, was built to facilitate detailed optimisation through collaboration with Designers, enable surface contamination testing and increase confidence in numerical vehicle drag predictions. This was constructed using a milled foam top-hat on a donor Range Rover underfloor. Inter-changeable modules for key areas, such as Air Curtain and Spoiler enabled efficient use of tunnel time.

A number of tests were conducted jointly with the Designer and Clay Modellers to optimise details such as the Rear Lamps and Spoiler Winglets. The following section details some of the features developed during this delivery phase.

Fig. 4. Photographs of the New Discovery "aerobuck" being tested at MIRA

2.2.1 Air Curtains

In the early design stage, the front corners were clearly identified as one the most promising areas for drag reduction with an overall potential of about 15 counts (1 count = 0.001 C_D). Improvement was realised through a combination of conventional shape optimisation and the development of an Air Curtain. The geometry was optimised as far as possible whilst staying true to the Design vision – a combination of increasing front wheel coverage, having a more vertical wheel arch ribbon and increasing the convexity of the profile in plan-view to get more suction locally. An Air Curtain was then introduced to further control the flow.

A comprehensive CFD study was conducted to characterise drag sensitivity to the key geometric parameters of the Air Curtain and understand subsequent changes to the flow. From this three concepts were developed and tested on the "Aerobuck" in both moving and static ground wind tunnels showing potential to reduce drag by up to 6 counts.

As seen in Fig. 5, the Air Curtain compensates for some of the lack of longitudinal flow momentum near the body side due to a large angle of flow separation at the front wheel arch liner, laterally kicking the flow away from the side panels.

2.2.2 Under-Floor

A key challenge for the New Discovery was developing an aerodynamic under-floor with best in class off-road capability and robustness. An aerodynamically efficient floor is usually achieved by maintaining as much longitudinal flow momentum as possible as well as managing flow separations. The aerodynamic floor strategy is presented in Fig. 6.

The engine and side under-trays were carried over from previous vehicles sharing this architecture. The large aluminium transmission under-tray has been clad in a smooth skin giving an incremental 3 drag count benefit and improving NVH performance. Its trailing edge has been tuned to slightly deflect the flow down to create a stable recirculation bubble at the break-over x-location, thus maintaining flow momentum. The flow re-attaches onto the exhaust SCR can at the vehicle centreline and onto the urea and fuel tank cradles as illustrated in Fig. 7(a). These protective shields have been extended to maximise coverage and profiled to ensure minimal stagnation pressure.

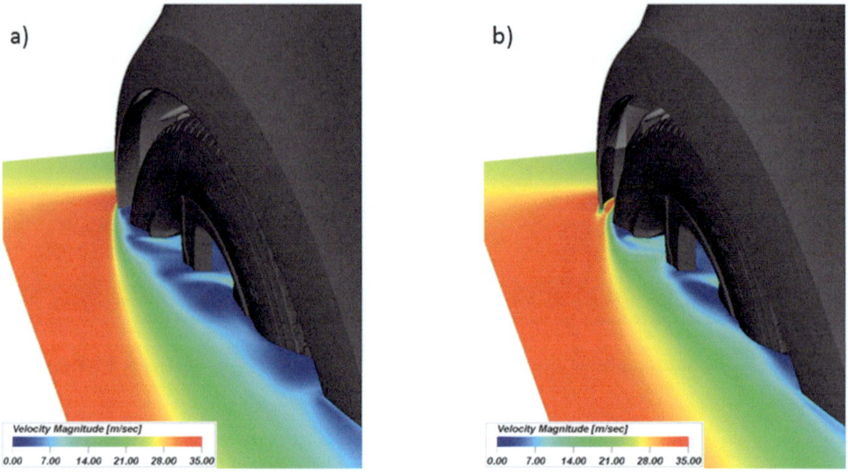

Fig. 5. Effect of air curtain on the flow field (Horizontal slice +150 mm from wheel centre): (a) without and (b) with air curtain

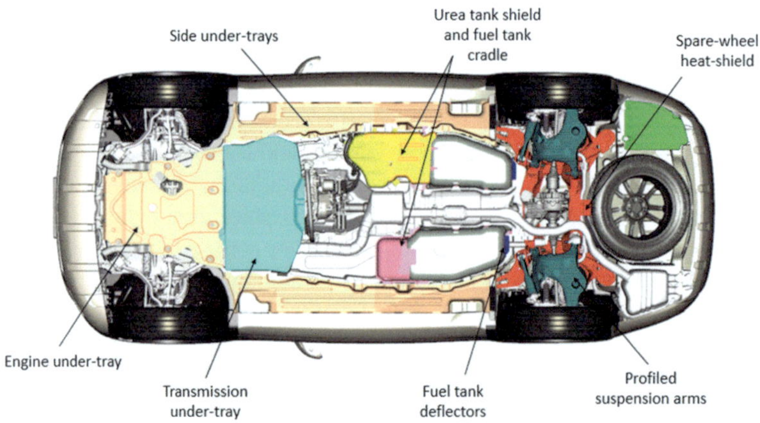

Fig. 6. Aerodynamic under-floor strategy

The large radius used on the lower trailing edge of the fuel tank cradle to avoid grounding was initially adverse for drag, causing the flow to turn upwards in an unsteady manner. Deflectors have been developed to guide the flow more efficiently and preventing it from impinging on the sub-frame and suspension arms. This resulted in a drag reduction of 2 counts.

The heat shield surrounding the spare wheel was optimised to give the required thermal protection without a drag penalty. For powertrains with a single sided exhaust muffler, an in-fill tray was engineered to bridge the void between the suspension and rear bumper.

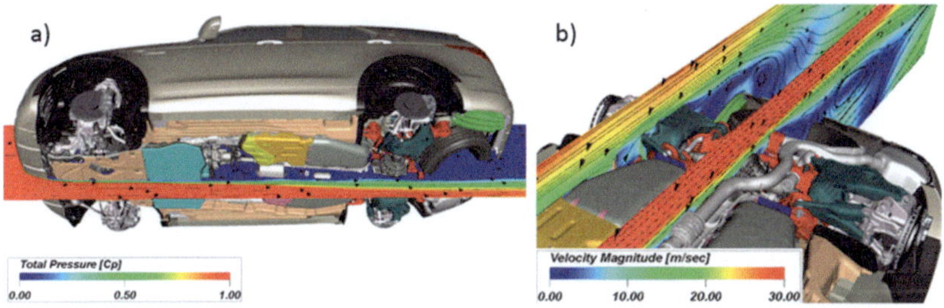

Fig. 7. CFD results illustrating the final flow field under the production vehicle: (a) total pressure coefficient vertical slice at vehicle centreline and (b) velocity magnitude vertical slices at centreline and offset by 0.5 m

Eventually, the development of a low-drag wake was achieved by forcing the flow to separate steadily over time along the y-axis, at the same x-location. At the vehicle centreline, the heat-shield z-location was optimised to ensure separation at this location, thus enabling the drag to be insensitive to the presence of the underslung spare wheel, or its size. Outboard of the centreline, the rear suspension arm trailing edge was used to force separation as illustrated in Fig. 7(b).

2.2.3 Slotted Roof Spoiler

Rear surface contamination is essentially an aerodynamically driven issue [14] and was identified as an area for improvement over the out-going model. To address this, a roof spoiler with slots creating a secondary flow tangent to the rear glass was developed. Previous studies on square back cars [15, 16] and trucks [17] have shown that a large reduction of rear glass soiling can be achieved using this technique, but usually at the expense of a significant drag coefficient increase.

Many different spoiler configurations were tested for both aerodynamics and soiling in the FKFS Aeroacoustic and Thermal wind tunnels respectively and in CFD. The greater the flow rate through the slots, the greater the reduction in rear soiling, but at high flow rates this results in a significant detriment to the drag coefficient as well as a large increase in rear lift. Configurations of slot were identified with a moderate flow rate which nevertheless enables an improvement of 15 to 30% in rear glass cleanliness and are neutral for drag, or even offer a small benefit.

Figure 8 shows the wake structure at the vehicle centreline as well as the static pressure coefficient (C_p) distribution over the base. Although the overall wake structure does not change significantly, the introduction of a span-wise vortex structure behind the rear glass can be seen. Even though the presence of this structure reduces the static pressure on the glass and increases the drag locally as seen in Fig. 9, the pressure recovery in the tailgate area is greater which cancels out the negative impact on the base drag.

Figure 9 highlights the subtle wake re-organisation causing the increase in pressure recovery over the tailgate. The span-wise vortex structure draws down the top shear layer slightly, reducing the height of the recirculation areas as well as providing a better condition for the side shear layers above and below the vehicle waistline, resulting in a

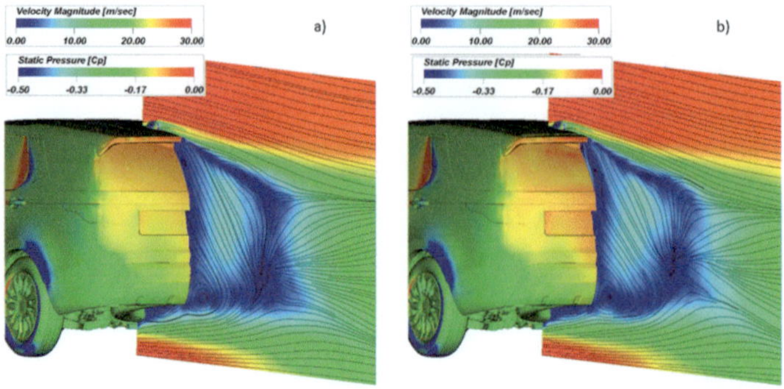

Fig. 8. Effect of the slotted spoiler on base pressure and velocity magnitude at vehicle centreline: (a) without and (b) with slotted spoiler

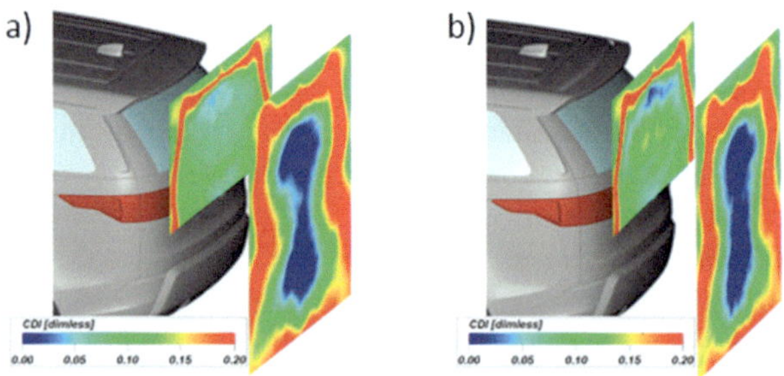

Fig. 9. Effect of the slotted spoiler on local drag coefficient: (a) without and (b) with slotted spoiler

less squeezed wake at mid-height. The local drag C_{Dl} (originally termed "micro-drag"), defined in Cogotti [18] and provided in Table 1 illustrates the increase in drag behind the rear glass with the slotted spoiler but also the resulting overall improvement downstream. This specific configuration resulted in a drag reduction of 3 counts. The rear glass soiling benefit seen during on-road testing was measured to be about 20%.

In addition to the spoiler, the winglets shown in Fig. 8 bring an additional benefit of 2 drag counts compared to the configuration without them, illustrated in Fig. 9.

2.2.4 Eco-Wheel

A CO_2 focused wheel and tyre combination was engineered for this programme, consisting of a narrow 235 mm (nominal) width, low rolling resistance tyre with an aerodynamically optimised 20″ wheel style. Two design themes were developed for the wheel using guidelines supplied by the Aerodynamics Team. A sensitivity study was conducted to guide the optimisation of the preferred theme. A summary of the configurations tested is shown in Table 2 along with their respective numerically predicted changes in drag coefficient C_D, the fan or ventilation moment $C_{D,VENT}$ [19–22] given in equivalent C_D and the brake cooling coefficient BCC. A scaled version of the 21″

Table 1. Local drag coefficient in the vehicle near wake for vehicle with and without slots

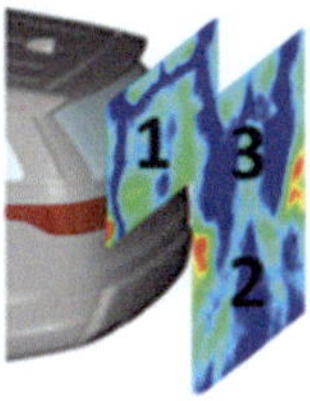

	Without slots			Slotted spoiler		
	C_{Dl}	C_{Dv}	C_{Dw}	C_{Dl}	C_{Dv}	C_{Dw}
Slice 1	**0.077**	0.005	0.072	**0.082**	0.005	0.077
Slice 2	**0.095**	0.006	0.089	**0.090**	0.006	0.084
Slice 3	**0.105**	0.005	0.100	**0.103**	0.004	0.099

Table 2. Influence of geometrical wheel changes on C_D, fan moment $C_{D,FM}$ and BCC

	BASELINE	1	2	3	4	5	6 / FINAL
WHEEL STYLE	Range Rover HEV wheel	Cyclone as received	As 1 with chamfers on spokes	As 1 with thinner spokes	As 1 without back star	Concave annular blanking	Convex annular blanking

ΔC_D	-	-0.006	-0.007	-0.006	-0.008	-0.005	-0.008
$\Delta C_{D,VENT}$	-	+0.001	+0.001	+0.001	+0.001	+0.001	+0.001
ΔBCC	-	-1BCC	-1BCC	0	0	-1BCC	-1BCC

wheel offered on Range Rover hybrids was used as a baseline – representing a well performing wheel.

Configuration 1 results in a drag benefit of 6 counts and shows a good Design interpretation of the aerodynamics guidelines, whilst not significantly degrading brake cooling. Configurations 2 and 3 show attempts to change the wheel fan moment which eventually remained within a (effective) drag count during the study.

The sensitivity study also identified annular blanking of the radial profile as the critical parameter for this wheel design, consistent with the findings of other workers [23]. Going from a concave profile to a convex profile laterally more flush to the outermost face of the rim induces more positive longitudinal entrainment of the flow near the ground, thus reducing the length of the lower structure originating from the flow separation around the tyre shoulder. This change of wheel wake structure observed in Fig. 10 unlocks an additional 3 drag counts with minimal impact on Design and feasibility.

Eventually style 6 with a black and diamond turn finish was chosen as the wheel style moving forward. This wheel presented in Fig. 11 reduces the drag by 8 counts.

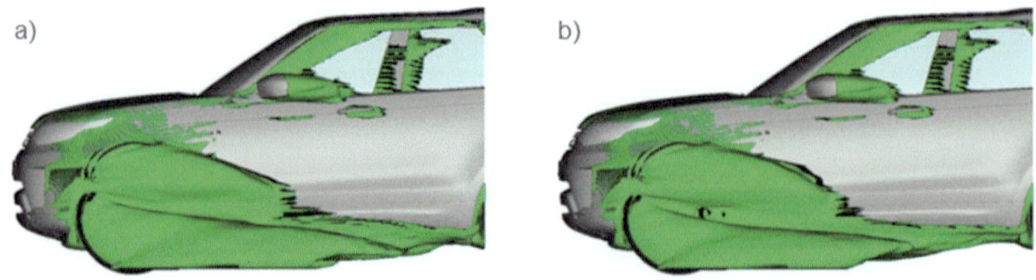

Fig. 10. Iso-surface of total pressure coefficient CpT = 0 for configuration: (a) 5 and (b) 6

Fig. 11. Aerodynamic "Eco-wheel" developed for the 17MY Discovery

2.2.5 Active Devices

To aid delivery of the aerodynamic target and help balance the other attributes of the vehicle, active devices have also been employed. The vehicle is offered with air suspension, which provides enhanced off-road capability and allows the car to automatically lower the ride height when cruising at motorway speeds. This reduces drag by 6 counts (uncorrected for frontal area change) at speeds where aerodynamic drag represents a significant proportion of the road load.

Active grille shutters are fitted to the upper cooling aperture allowing cooling airflow to be optimised to the driving conditions.

2.3 Validation and Correlation

Testing of validation prototype vehicles was done predominately in the FSWT for consistency with development, however some exploratory testing was also conducted in the AAWT to aid understanding in preparation for the upcoming changes in certification procedures. This confirmed that a drag coefficient of $C_D = 0.33$ had been achieved in both wind tunnels – the lowest to-date for a Land Rover and a 17% improvement over the outgoing model.

Although previous experience with the FSWT has shown good correlation between numerical and experimental results for SUVs, usually within ± 5 counts for the predicted drag coefficient, the increased complexity due to modelling rotating wheels as well as the flow structures associated with the tyre tread and contact patch highlighted the need to correlate the tools with the new methods.

Fig. 12. New Land Rover Discovery VP at FKFS wind tunnel (left) and CFD model as per FKFS configuration (right)

A correlation exercise was carried out on a validation prototype vehicle equipped with a 2.0 I4 Diesel engine and the 20″ Aerodynamic wheel discussed previously fitted with 235/60 Continental Cross Contact tyres (Fig. 12). The CFD runs were specified with a low-blockage (0.1%) domain and a moving ground plane. Rotating velocity boundary conditions were applied to de-featured axisymmetric tyres whilst the rim and spokes were rotated using a sliding mesh (LRF) zone. Unfortunately, this setup does not capture the geometrical contact patch and the loaded tyre deformation as the tyre intersects with the ground. Also, a slight difference of 5 mm in ride heights between the CFD model and the vehicle tested in the tunnel was corrected for by adding 2 counts to the physically measured value.

The absolute drag coefficients for this vehicle in two different cooling configurations are given in Table 3. Blanking all the front apertures results in a drag reduction of 15 counts in the wind tunnel; however, the drag delta predicted in CFD is about 1 count only. Predicting the cooling drag delta is known to be a difficult task [24] as it involves many flow field interactions around the vehicle [25] and this lack of correlation was expected.

Table 3. Cooling drag deltas of the prototype measured in the AAWT and predicted in CFD

Vehicle configuration	AAWT	CFD
Cooling drag delta	0.015	0.001

Using the 2-gradient method for open jet wind tunnels of Mercker and Cooper [26, 27], a correction of +12 drag counts was applied to the measured drag to account for the pressure gradient and the blockage in the test section. Moreover, the additional aerodynamic drag of the restraining struts increases the overall measured value by 2 counts [28], bringing the total wind tunnel correction to +10 counts. Figure 13 illustrates the different corrections applied to the raw experimental data and highlights an excellent agreement - within 2 drag counts between the absolute drag values predicted by the two tools when the cooling apertures are open.

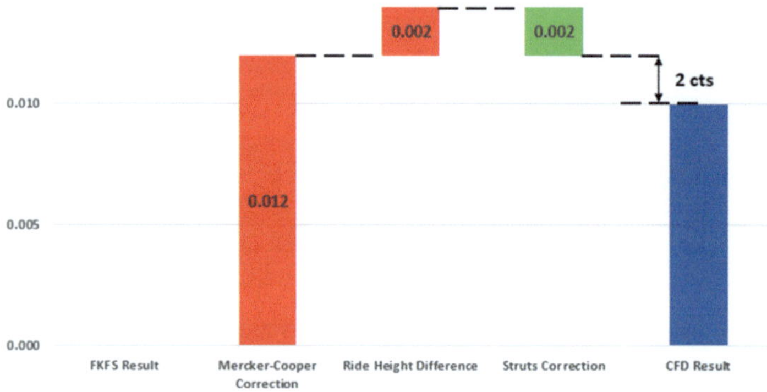

Fig. 13. Wind tunnel corrections for the vehicle prototype with all cooling apertures open

Obviously, the closed cooling configuration does not correlate as well. A comparison of some of the drag coefficient deltas induced by a number of configuration changes is given in Fig. 14. Overall, the trends when modifying the under-floor are well captured by CFD although the drag delta magnitudes are not in perfect agreement. Removing the spare wheel resulted in a benefit of 1 drag count measured in the wind tunnel and almost 2 counts in CFD. This is an interesting result as previous tests in the FSWT and in CFD using a static ground boundary condition had shown no sensitivity in this area. This difference is believed to be due to the increase of momentum under the vehicle in moving ground configuration which makes the flow partially reattach onto the spare-wheel. This is illustrated in Fig. 15, which shows the difference in static pressure predicted by CFD at the centreline between static and moving ground configurations for the same vehicle geometry. This mechanism also lowers the pressure behind the rear bumper, leading to the small drag detriment.

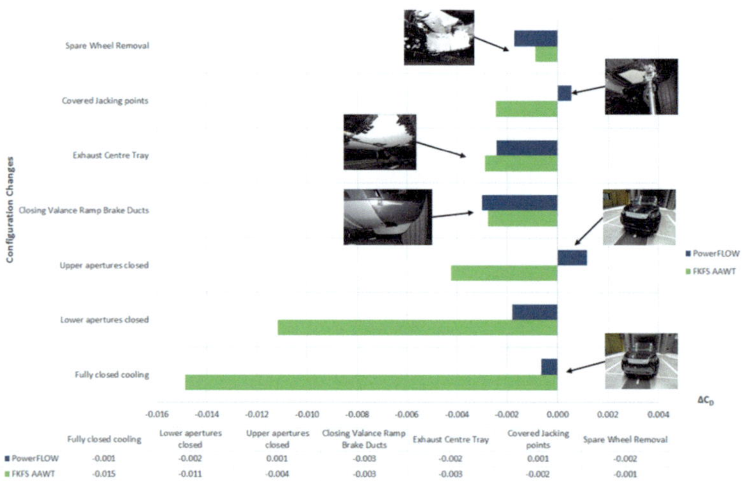

Fig. 14. Comparison of the experimental and numerical drag coefficient deltas induced by configuration changes

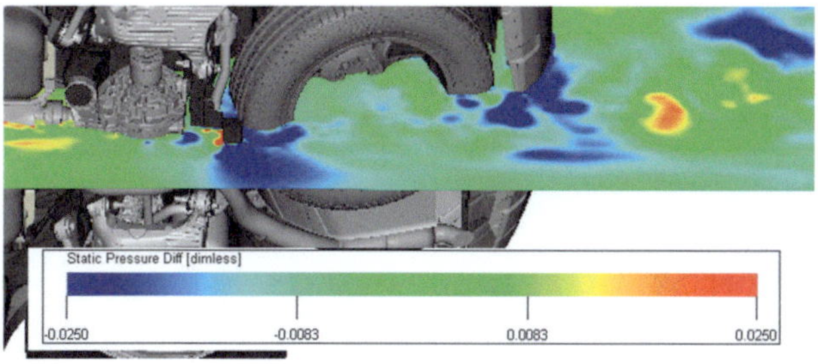

Fig. 15. CFD prediction of the static pressure coefficient difference between static and moving ground boundary conditions for the same vehicle geometry

The trend is predicted incorrectly when covering the jacking points, as CFD predicts a small drag detriment whereas a 2 count benefit is measured in the wind tunnel. The reason for this discrepancy is not fully understood but plausible explanations may be the presence of the restraining struts in the tunnel which may be affecting the flow over the jacking points, or the tyre modelling simplifications that might result in wheel wake differences. Blanking the brake cooling valance ramps resulted in a drag reduction of 3 counts with both tools.

Some alternative cooling configurations have also been compared by blanking the upper or lower cooling apertures. As for the fully closed cooling configuration, the different drag benefits measured in the wind tunnel are not agreement with the CFD predictions. The predicted trend is even incorrect: when blanking the upper aperture CFD predicts an increase in drag. These results again highlight the difficulties in predicting cooling drag and the importance of investigating this problem to get a better understanding on it.

This correlation exercise in moving ground rotating wheel conditions for a large SUV was the first of its kind within Jaguar Land Rover and shows a good agreement between our new CFD approach and the experiments in the AAWT. It also validates the aerodynamic development process for future vehicles which involve an exclusive usage of CFD in the early design phase, until a physical property is available.

3 Summary

Some elements of the aerodynamic development process of the New Land Rover Discovery have been described. To deliver the ambitious aerodynamic drag target, the out-going model underwent a spectacular transformation, enabling the drag coefficient to be reduced by 70 counts, down to $C_D = 0.33$, the lowest figure to date for a Land Rover SUV (Fig. 16).

To enable such a drag reduction, an optimisation of the main vehicle proportions, combined with some new elements of the under-body and an aerodynamic wheel were needed. A number of features reducing the overall drag further have also been presented.

Fig. 16. CFD rendering of the J#1 CFD model of the new Land Rover Discovery

Throughout its development, the Land Rover Discovery has seen some important changes in processes and best practices to prepare for the future certification procedures, moving from the static ground wind tunnel of MIRA to the moving-ground FKFS wind tunnel and using new CFD boundary conditions with moving ground rotating wheels.

A correlation study has shown a good agreement between CFD and FKFS drag measurements, once wind tunnel corrections are applied. Apart from the cooling drag prediction, the trends when making configuration changes were also identical with similar magnitudes.

Acknowledgements. The authors would like to acknowledge the support from Aerodynamics Team at Jaguar Land Rover, Gaydon, UK. The authors would also like to thank Jaguar Land Rover for giving their permission to publish this paper.

References

1. Howell, J., Gaylard, A.: Improving SUV aerodynamics. In: 6th MIRA International Vehicle Aerodynamics Conference, pp. 1–17 (2006)
2. Gaylard, A.P.: The Appropriate Use of CFD in the Automotive Design Process. SAE Technical Paper 2009-01-1162 (2009)
3. Samples, M., Gaylard, A.P., Windsor, S.: The aerodynamics development of the Range Rover Evoque. In: 8th MIRA International Conference on Vehicle Aerodynamics (2010)
4. Gaylard, A.: The new Jaguar XE: a case study in computational aerodynamics. In: International Forum Advanced Automotive Aerodynamics (2015)
5. Chen, H., Chen, S., Matthaeus, W.H.: Recovery of the Navier-Stokes equations using a lattice-gas Boltzmann method. Phys. Rev. A **45**(8), R5339–R5342 (1992)
6. Chen, H., Teixeira, C., Molvig, K.: Digital physics approach to computational fluid dynamics: some basic theoretical features. Int. J. Mod. Phys. C **8**(4), 675–684 (1997)
7. Chen, H., Kandasamy, S., Orszag, S., Shock, R., et al.: Extended Boltzmann kinetic equation for turbulent flows. Science **301**(5633), 633–636 (2003)
8. Gaylard, A.P.: The aerodynamic development of the new Jaguar XF. In: 7th MIRA International Conference on Vehicle Aerodynamics (2008)

9. Gaylard, A., Pitman, J., Jilesen, J., Gagliardi, A., et al.: Insights into rear surface contamination using simulation of road spray and aerodynamics. SAE Int. J. Passeng. Cars Mech. Syst. **7**(2), 673–681 (2014)
10. Newnham, P., Passmore, M., Howell, J., Baxendale, A.: On the Optimisation of Road Vehicle Leading Edge Radius in Varying Levels of Freestream Turbulence, SAE Technical Paper 2006-01-1029 (2006)
11. Carr, G., Stapleford, W.: Blockage Effects in Automotive Wind-Tunnel Testing, SAE Technical Paper 860093 (1986)
12. Künstner, R., Potthoff, J., Essers, U.: The Aero-Acoustic Wind Tunnel of Stuttgart University, SAE Technical Paper 950625 (1995)
13. Blumrich, R., Widdecke, N., Wiedemann, J., Michelbach, A., et al.: New FKFS technology at the full-scale aeroacoustic wind tunnel of University of Stuttgart. SAE Int. J. Passeng. Cars Mech. Syst. **8**(1), 294–305 (2015)
14. Gaylard, A.P., Kirwan, K., Lockerby, D.A.: Surface contamination of cars: a review. Proc. Inst. Mech. Eng. Part D J. Automobile Eng. **231**(9) 1160–1176 (2017)
15. Janson, J., Darrieutort, L., Bannister, M.: New development and working methods used in the aerodynamic development of the new large estate, JSAE paper 20005352 (2000)
16. Costelli, A.F.: Aerodynamic characteristics of the Fiat UNO car, SAE paper 840297 (1984)
17. Fujimoto, T., Miyake, N., Watanabe, Y.: Suppression of mud adhesion to the rear surface of a van-type truck, SAE paper 920203 (1992)
18. Cogotti, A.: A Strategy for Optimum Surveys of Passenger-Car Flow Fields, SAE Technical Paper 890374 (1989)
19. Wickern, G., Zwicker, K., Pfadenhauer, M.: Rotating Wheels - Their Impact on Wind Tunnel Test Techniques and on Vehicle Drag Results, SAE Technical Paper 970133 (1997)
20. Landström, C., Josefsson, L., Walker, T., Löfdahl, L.: An experimental investigation of wheel design parameters with respect to aerodynamic drag. In: 8th FKFS Conference (2011)
21. Modlinger, F., Demuth, R., Adams, N.: New directions in the optimization of the flow around wheels and wheel arches. In: MIRA International Conference on Vehicle Aerodynamics (2008)
22. Vdovin, A., Lofdahl, L., Sebben, S.: Investigation of wheel aerodynamic resistance of passenger cars. SAE Int. J. Passeng. Cars Mech. Syst. **7**(2), 639–645 (2014)
23. Landström, C., Walker, T., Christoffersen, L., Löfdahl, L.: Influences of Different Front and Rear Wheel Designs on Aerodynamic Drag of a Sedan Type Passenger Car, SAE Technical Paper 2011-01-0165 (2011)
24. Larson, L., Gin, R., Lietz, R.: Aerodynamic investigation of cooling drag of a production sedan Part 2: CFD results. SAE Int. J. Passeng. Cars Mech. Syst. **10**(1), 628–637 (2017)
25. Pitman, J.: An experimental investigation into the flow mechanism around an SUV in open and closed cooling air conditions. In: FKFS – 11th Aerodynamic Conference (2017)
26. Mercker, E., Cooper, K.: A Two-Measurement Correction for the Effect of a Pressure Gradient on Automotive, Open-Jet, Wind Tunnel Measurements. SAE Paper 2006-01.0568 (2006)
27. Mercker, E.: On buoyancy and wake distortion in test sections of automotive wind tunnels. In: FKFS – 9th Aerodynamic Conference, pp. 205–227 (2013)
28. Wiedemann, J., Potthoff, J.: The New 5-Belt Road Simulation System of the IVK Wind Tunnels - Design and First Results SAE Technical Paper 2003-01-0429 (2003)

The Aerodynamics Development of the New Volkswagen Polo

Carsten Repmann and Mathias Hähnel(✉)

Volkswagen AG, Berliner Ring 2, 38436 Wolfsburg, Germany
Mathias.Haehnel@volkswagen.de

Abstract. The Polo is entering its 7th generation. The car is based on a new architecture concept using a new modular design. Both the car body and the platform are newly designed. The platform will be used by several brands of the Volkswagen group for a variety of vehicles.

The Polo has been designed to meet demanding emission requirements. One of the key contributors to achieving the requirements is improved aerodynamics in spite of a significant growth of vehicle frontal area. These improvements were applied to the entire fleet of vehicles, thus ensuring that future legislative targets will be met.

1 Introduction of the New Polo

1.1 Evolution from Predecessor

The new Polo is based on a new modular. This approach facilitates the use of certain components from larger vehicle classes. However, it also leads to an increase in vehicle dimensions compared to the Polo's predecessor. Frontal area grows from 2.04 m^2 to 2.12 m^2 (Fig. 1).

Fig. 1. Comparison of frontal area of the new Polo to its predecessor (outline)

J. Wiedemann (ed.), *Progress in Vehicle Aerodynamics and Thermal Management*,
https://doi.org/10.1007/978-3-319-67822-1_10

Increased size also shows in details like the minimum wheel dimensions. From the perspective of aerodynamics the development of the new Polo started with an extra burden in the guise of increased component and overall vehicle sizes (Table 1).

Table 1. Characteristic measurements of wheels

	Previous	New
Wheel diameter	610 mm	635 mm
Min. wheel width	175 mm	185 mm

1.2 Interior Dimensions

The growth of the new Polo offers improvements in interior dimensions for customers. However, in order to support aerodynamics requirements an agreement with the vehicle layout department was struck regarding roof angle and cabin width at the rear of the vehicle. The compromise involved tapering of the rear end. The impact on headroom was contained by changes to the rear seating position. This way most of the rear seat head room originally envisaged could be retained (Fig. 2).

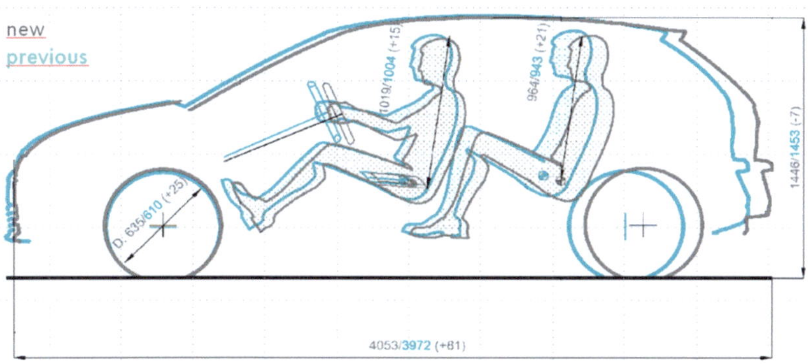

Fig. 2. Characteristic measures and seating layout

2 Development of Platform and Car Body

The new Polo is not only a new vehicle generation, it also uses a new platform.

For most of the bigger vehicles in VW's portfolio a switch had already been made from a platform based approach to an even more general modular design for the vehicle.

This approach was transferred into the small vehicle (a.k.a. Polo) class.

The 7th generation of Polo therefore now has significant similarity if not commonality in architecture and parts with larger vehicles like the Golf and Passat.

2.1 Platform

The Polo is based on a new modular design which is the basis for several cars from each of the brands:

- Audi
- Seat
- Skoda
- Volkswagen

It also covers various vehicle lengths and overhangs.

The first step therefore was the development of the modular design and thus an underbody suiting the needs of the different users.

2.1.1 Facilities Used for Development/Cooperation with Audi

The modular design will be the basis for a large range of vehicles for the next ten years. Therefore it has to cover the needs of both present and future vehicle designs. The development of the platform was the responsibility of Volkswagen, however taking into account the requirements of the other brands which will use the platform for their projects. In order to achieve a state-of-the-art design, wind tunnels with ground simulation were used for platform development. Audi was very supportive by making their wind tunnel available. Further work was performed at FKFS with CFD providing its share as well. Overall about 200 h of wind tunnel time were devoted to platform development.

In contrast to the approach for the platform, all of the roughly 450 h of wind tunnel testing for the development of the car body was done at the time-honoured Volkswagen wind tunnel in Wolfsburg (Fig. 3).

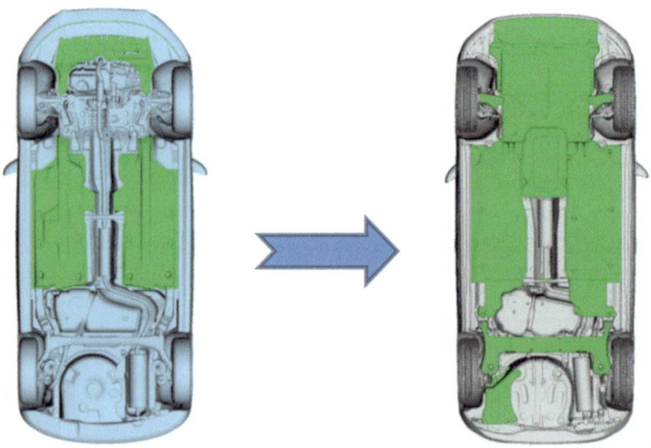

Fig. 3. New Polo underbody concept

2.1.2 Underbody Concept

Even though the Polo belongs to the range of small vehicles that are very cost sensitive, a decision was made to use a rather comprehensive set of underbody panels. At the time this approach was unprecedented in the small vehicle segment. The previous generation of Polo was equipped with a basic set of underbody panels which could already be considered above average for the small segment. The underbody panels were supported by an air dam below the front bumper.

Because of the extensive paneling and smoothing of the underbody of the new generation Polo the requirement for the air dam was dropped. The panels are used for almost all vehicles, the exceptions being those which are either restricted by thermal requirements or special package needs e.g. for CNG powered cars.

The underbody panels are smaller and more numerous than can be seen on some competitor cars. This is due to the requirement of suitability for different track widths, wheel bases and drive concepts all of which are supported by the modular design.

2.1.3 Underbody Details

In most present VW cars brake cooling is supported by brake cooling ducts that channel air from below the frontend over the wheel deflectors into the wheel houses. Unfortunately from an aerodynamics perspective this increased cooling air mass flow into the wheel houses carries a drag penalty. Therefore a new type of brake cooling duct was designed in cooperation with the chassis department. This new duct uses a smaller mass flow and guides the air more precisely to the caliper. Moreover the wheel deflectors are made from a softer material so that a reduced ground clearance of the deflectors becomes possible (Fig. 4).

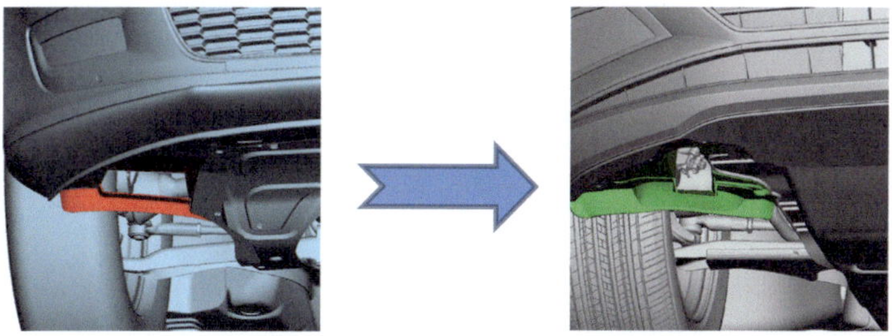

Fig. 4. Development of brake cooling duct

An interesting observation was made with the fairing of the rear axle. The first design had a drooped trailing edge. In wind tunnel tests an unusually pronounced sensitivity to ride height changes was observed. Eventually, this could be attributed to the shape of the fairing's trailing edge. The adverse effect of the kick down only became apparent at ride heights other than the default height. With a straight trailing edge the drag sensitivity could be reduced to a normal level (Fig. 5).

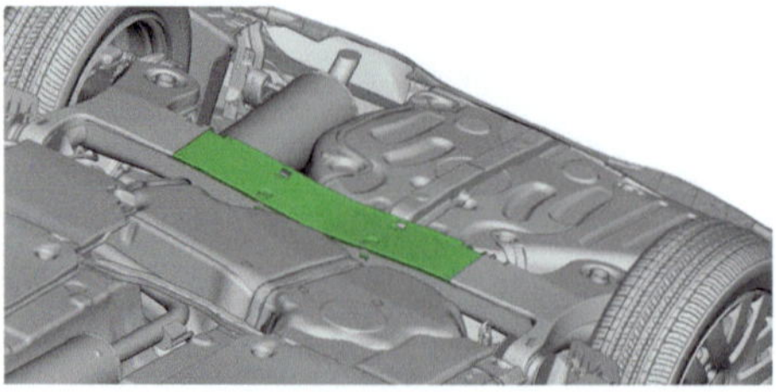

Fig. 5. Rear axle fairing with drooped trailing edge

To improve acoustic properties, a stiffener is welded to the spare wheel well. At first this component was attached in a transverse position. In terms of ground clearance and visibility this is the preferred position. However, after an intervention on behalf of aerodynamics a longitudinal mounting has been chosen (Fig. 6).

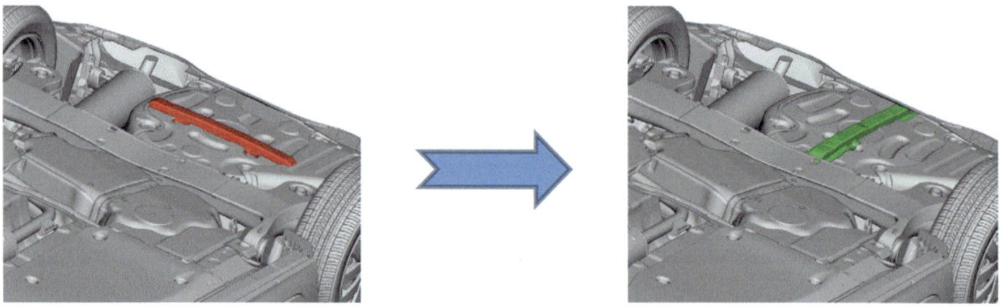

Fig. 6. Changing position of the acoustics stiffener under the spare wheel well

2.2 Upper Body

2.2.1 Layout Framework

The modular design approach dictated an increase in vehicle width by more than 60 mm.

For the interior dimensions an early agreement was found with the vehicle layout department addressing aerodynamics concerns. This made the following changes to the original layout possible:

- Increased slant angle of the roof
- Increased tapering of the lateral rear end
- Reduced vehicle base area

These measures were meant to contain the negative aerodynamic effects from the increased overall width. They were used as a starting point for the styling department.

2.2.2 Geometric Details

To improve the basic shape of the car a slightly lower inclination of the windscreen and a more upright backlight angle were chosen. However, these changes were clearly

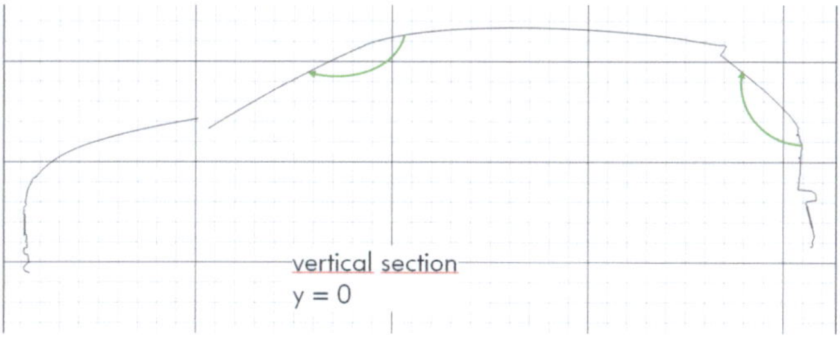

Fig. 7. Inclination of windscreen and backlight

limited by requirements for the optical properties of the windscreen as well as styling expectations for the backlight (Fig. 7).

On the previous Polo a drain strip had been added to the A-pillar for the low drag Blue Motion derivative. This device reduced drag but incurred the cost of extra components. In the early stages of development it was assumed that a drain strip would also be needed for the new Polo to meet its challenging aerodynamics targets. However, at the same time an alternative approach was pursued. A new shape for the A-pillar was designed with a reduced height of the step from the windscreen to the crest of the pillar. This new shape delivered a drag benefit while at the same time satisfying the requirements for aeroacoustics and side glass soiling. Unfortunately a very prominent design feature was introduced to the A-pillar—a kink spoiling the flow. If it wasn't for this particular design, the drag caused by the A-pillar could be even lower (Fig. 8).

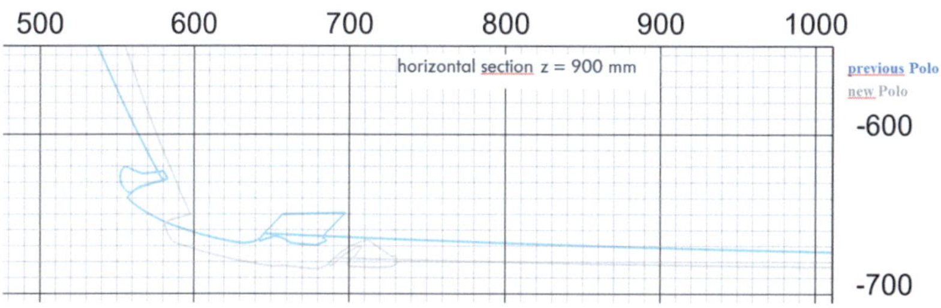

Fig. 8. Section of the A-pillars

At the rear of the car lateral interior space was reduced below initial requirements to allow for rear end tapering. This change to the vehicle layout had been identified to facilitate a significant improvement of overall drag. However, the aerodynamic benefits of the tapering could not be fully reaped: The selected design has a kink at the end of the C-pillar. The kink causes premature flow separation on the C-post leading to a double drag penalty because any possible additional C-pillar finlets are ineffective in the separated flow downstream of the kink (Figs. 9 and 10).

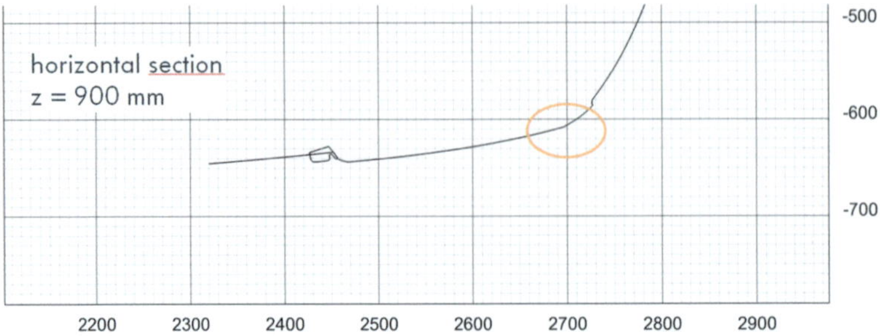

Fig. 9. Horizontal section of the C-pillar at z = 900 mm

Fig. 10. Horizontal section of the flow field

3 Whole Vehicle

3.1 Sensitivity to Ground Simulation

Previous generations of the Polo were developed without the use of ground simulation (i.e. with the Wolfsburg wind tunnel). These vehicles tend to experience increased drag in wind tunnels with ground simulation.

The new Polo is the first car in the VW hatchback fleet that improves with ground simulation. The resulting drag reduction is about 4%.

3.2 Fleet-Wide Drag Reduction

Instead of designing a special version of the car with particularly low drag the emphasis was on benefits for the entire fleet. Therefore the spread of drag coefficients for different versions of the car is rather small. Care was taken to produce a car that addresses everyday situations, e.g. typical loading conditions. For the design ride height the vehicle was loaded with 150 kg instead of the EADE load in order to be closer to a typical customer use case. This load was also used for designing the underbody panels: They have the best performance for a lightly loaded car because this is assumed to be the primary operating condition of the Polo (Fig. 11).

Fig. 11. Ranges and average numbers for drag coefficient

3.3 Aerodynamic Properties

The new Polo achieves $c_d = 0.298$. Compared to its predecessor this is a reduction by 8%. Drag area is still reduced by 4% despite the increased dimensions of the car. The expected fleet average for the new Polo is $c_d = 0.306$.

These numbers are achieved without the use of active elements like grill shutters (Fig. 12).

Fig. 12. The new Polo in the wind tunnel

4 Conclusions

- The new Polo has significantly increased in size
- Basis for the new Polo is a new modular design
- A comprehensive underbody concept has been developed and applied
- Numerous aerodynamic improvements have been made to the upper body of the Polo
- Despite the increased size of the car the drag area $c_d \cdot A$ is 4% smaller than the predecessor's
- The range of drag coefficients starts at $c_d = 0.298$ with an expected fleet average of $c_d = 0.306$

On-Road Tests

Aerodynamic Development of a New Coach Generation Based on Wind Tunnel Testing, CFD-Simulation and On Road Tests

Marius Hellmold(✉), Stephan Kopp, Andreas Liebing, and Stephan Schönherr

MAN Truck & Bus AG, Dachauer Str. 667, 80995 Munich, Germany
Marius.Hellmold.a@man.eu

Abstract. In order to significantly reduce the vehicle's fuel consumption, a lot of attention was paid to the new design of the NEOPLAN TOURLINER. During the design phase, each detail was therefore looked at closely in order to determine how to implement the NEOPLAN design criteria, specific vehicle design and detailed design in a way that optimizes the drag value coefficient (c_D). Through optimization loops, we found a target-oriented design using CFD and implemented this meticulously in the area at the front of the bus, the front roof dome, the mirrors, the air-conditioning system cladding, the wheel arch panels, wheel trim, entire rear panel, as well as areas in the underbody. The results were then examined on a scale model in the wind tunnel of Pininfarina. A lot of effort was put to build a model which includes finest details, e.g. the entire underbody-structure and (motorized) rotating wheels. Eventually, the drag value was confirmed through test drives of initial vehicles. These are constant speed tests in which probe data of drive momentum, vehicle speed and wind velocity/direction are used to calculate the c_D value.

The c_D value was reduced by over 20% in comparison to the Tourliner, released in 2002, from 0.46 to 0.36, now. The vehicle therefore achieves overall fuel savings of 10% in comparison to the predecessor bus, reducing fuel consumption to an average of approx. 20 L/100 km.

This represents a high degree of innovation, as the improved consumption is in a standard, entry-level vehicle of the NEOPLAN brand. The standard Tourliner not only has a more attractive acquisition cost – it is also designed to reduce consumption. More for people and the environment!

1 MAN Truck & Bus AG Welcomes a New, Very Efficient Member to the Family with the NEOPLAN Tourliner

The new NEOPLAN Tourliner is a true all-rounder and creates the opportunity to break into the premium segment of coaches. The key advantages for the customer are:

- Aerodynamic value improved by more than 20%. Fuel consumption reduced by 10%.
- Coach from the standard segment with the typical, highly appealing NEOPLAN design including the new ring-shaped LED daytime driving lights.

J. Wiedemann (ed.), *Progress in Vehicle Aerodynamics and Thermal Management*,
https://doi.org/10.1007/978-3-319-67822-1_11

- Flat floor enables flexible design of the interior for various uses.
- The new MAN D2676 LOH engine with optimized driveline and new MAN EfficientCruise with EfficientRoll significantly reduce consumption.
- Maximum safety thanks to numerous safety systems and compliance with directive ECE R66.02 on the strength of superstructures.
- Modular design: From the end of 2017, the coach will be available in four different lengths with two or three axles (Figs. 1 and 2).

Fig. 1. The new NEOPLAN Tourliner

Fig. 2. The NEOPLAN family with new Tourliner

2 Economy and Efficiency Through Aerodynamic Exterior Design

This bus impresses in both appearance and efficiency. The new exterior design has improved aerodynamic values once again by more than 20% in comparison to the predecessor model. The c_D value of the new NEOPLAN Tourliner is just 0.36, which is comparable to that of a modern middle-class van. A great deal of effort has gone into the development of the complex aerodynamics and flow optimization of the Tourliner. Therefore, a heavy focus was placed on the optimization potential, preliminary investigations and current condition analysis of the predecessor vehicle. The identified potential for improvement was used to create a work catalogue of the body parts to be optimized. When designing, implementing and transferring the NEOPLAN design criteria to the vehicle package of the new Tourliner, each and every external area of the vehicle is optimized to improve aerodynamics. Many details on the bus exterior can contribute to this. For example, the new exterior mirrors, which have less flow separation as well as a better view (Fig. 3). Also, the front panel with the relevant areas of the front-end corners is causing only minor flow losses (Fig. 4).

Fig. 3. The new mirror design and its flow separation

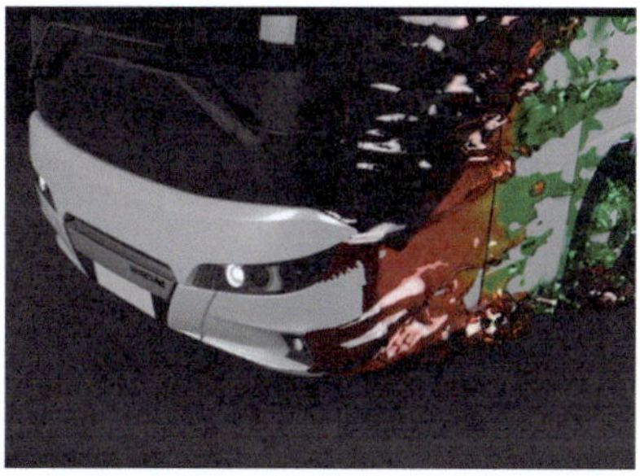

Fig. 4. Front/Sidewall transition with slight losses

And so are the corner radii of the windscreen with the associated rain channels integrated in the A-pillars, as well as the front roof dome at the top with the connected air-conditioning system, which is integrated on the roof in a smooth and appealing manner (Fig. 5). The lower front-spoiler edge generates a flow separation, which prevents the potential flow from colliding with the underbody parts, especially the front axle (Fig. 6).

Of course, the rear section and rear panel are also a major influencing factor in the aerodynamics of any bus or coach. It is here, that a significant share of the aerodynamics optimization can be achieved. Great care has been taken in the design of the rear panel with corresponding upper spoiler edge and lateral truncated rear end (Fig. 7). Here, the exterior components are implemented in small steps and optimization loops. Once the initial design CAS data has been created, this is checked using EXA

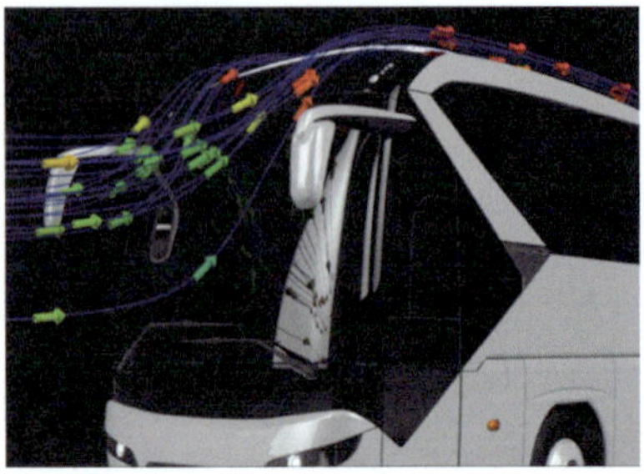

Fig. 5. Big roof dome radius leads to a smooth airflow

Fig. 6. Intended flow separation prevents the potential flow from colliding with axle components

PowerFlow. The resulting improvement potential then flows into the revision loop, and is implemented while still maintaining the balance between compliance with the technical vehicle specifications and dimensions, the vehicle package, the typical NEOPLAN design criteria and the best possible aerodynamics. MAN Truck & Bus AG has gone to great lengths and drawn on all its expertise to determine and validate the final result of the flow-optimized outer shell of the new NEOPLAN Tourliner, as well as precisely identify the c_D value of the vehicle. This has involved tests in the wind tunnel with a large scale model (1:2.5), driving tests on accompanying test routes, and motorway journeys with initial vehicles to obtain real data.

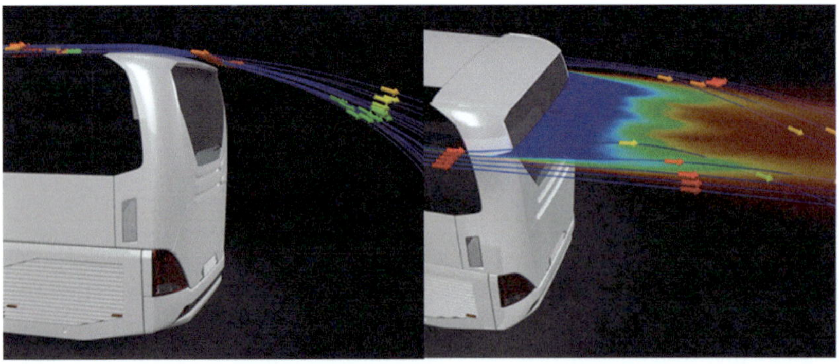

Fig. 7. Tail retraction at rear end reduces area at the base

3 Wind Tunnel Measurements

Besides the CFD-calculations, a wind-tunnel model was built with the scale of 1:2.5 (Fig. 8). As test facility, the wind tunnel of Pininfarina in Torino, Italy, was chosen. Several test sessions were used to both, redefine the actual c_D-value and further improve the vehicle's shape, especially at the underbody. To achieve profound results, MAN

Fig. 8. Scaled model (1:2.5) mounted on the 5-belt test section of Pininfarina wind tunnel

Fig. 9. Underbody structure, all built with in-house 3D-printers

expended a large amount of effort in building a highly detailed model. So, all underbody parts were reconstructed and built with the three in-house 3D-printers (Fig. 9). Basically, the model consists of a framework of aluminum profiles which is cased by wooden panels. The exterior design is cut out of ordinary modelmaking foam which is painted twice afterwards, to generate a smooth surface. The wind tunnel is equipped with a five-belt ground simulation and since the main belt can't be used to drive the wheels, all four tires contain electric motors. Their rotational speed is coupled with the moving-belt velocity, so turning on and off again the motor isn't bothering you.

The test session finally revealed a basic drag value of $c_D = 0.36$, a very neat result within the realms of commercial vehicles. Additionally, further aerodynamic improvements could be found.

4 Constant Speed Tests

In constant speed tests, drag-value determination is based on the assumption, that rolling resistance is constant with varying driving speeds.

At first step, the rolling resistance, or more specific the rolling wheel torque, must be detected at a low driving speed (15 km/h). This value can be subtracted from the driving resistance at high speed (89 km/h), which leads to the air resistance (Fig. 10).

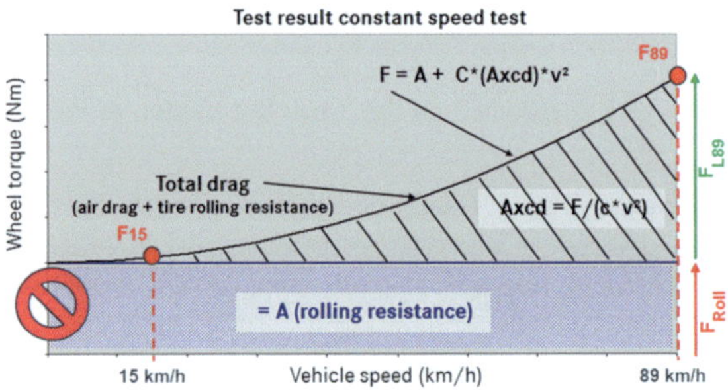

Fig. 10. Assumed composition of driving resistance as a function of vehicle speed

As test setup, the vehicle is equipped with torque measuring devices at the powertrain (e.g. hub-torque-devices, strain gauges, tire-torque-devices). Simultaneously, the actual flow velocity and direction is measured via an anemometer (Fig. 11), which must be placed far enough apart from the roof dome to ensure free-stream-velocity. However, if there is wind during the tests (as it's always the case), the measured velocities must be corrected to compensate for the velocity profile of the ground boundary-layer. And since you always get a geometrical misalignment of anemometer and vehicle x-axis due to installation, an angular correction is applied as well. Further corrections are also made regarding the influence of inertia forces [1].

The actual test drives are done on a test track (see [1] for track requirements) and at night, when wind speed is normally lower. At first, the vehicle must be brought to operating temperature during a 90 min round trip. The drive is then stopped at a suitable location, the drive axis is raised, and the signals from the torque measuring device are set to zero. This is followed by another warm-up of one round. The speeds and torques must have stabilized approximately 100 m before entering the 250 m measuring section, which has to be passed through with 15 km/h (lowspeed). The measurement is precisely started and terminated via dGPS. There are at least 20 passes

Fig. 11. The test object with applied anemometer

of the same measuring section with 89 km/h (highspeed), afterwards. Then, analogously to the first, a second set of lowspeed-tests is performed. At last, the vehicle is stopped at the appropriate position and the drift of the torque measurement signal is recorded again (Fig. 12).

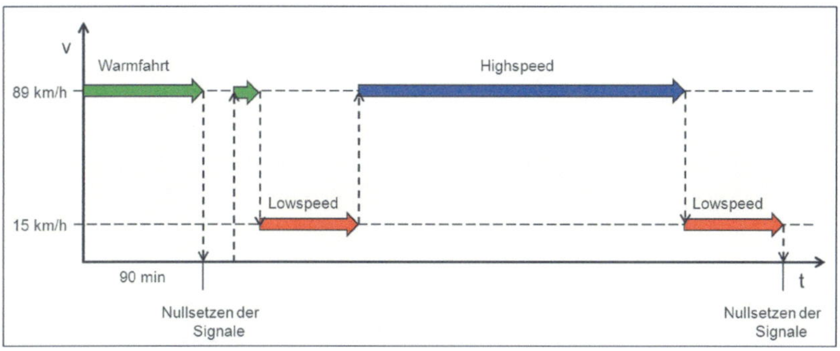

Fig. 12. Schematic test procedure for a single run

The test procedure is relatively simple in its execution. However, the calculated c_D values spread by approx. 10% and deviate on average about 12% from the values from the wind tunnel and the CFD calculations.

Some reasons of concern should be mentioned here. At first, assuming that rolling resistance is constant with different speeds is questionable. Influencing variables are, for example, loading and its distribution, tire air pressures and tire temperatures and even their speed itself. Here, even more investigations are necessary.

Further investigations should be also done on the ventilation moment of tires. For now, the calculated c_D values also include this additional air resistance, which actually must be subtracted.

5 Even More Efficiency

Today, optimized aerodynamics alone is not enough. From a technical point of view, the D2676 LOH engine also offers improved performance and torque, featuring an optimized driveline that allows for fuel savings of up to 10%. The long axle ratio of i = 2.73 and the optimized shifting strategy contribute to this, among other factors. Also available are the new generation of MAN EfficientCruise, and for the first time in coaches, the MAN EfficientRoll freewheel function. This "coasting" function delivers added efficiency even on slightly downhill gradients by automatically switching the gearbox to neutral position "N" and allowing the bus to roll with the lowest possible driveline friction loss.

Reference

1. AnnexVI - Regulation (EU) No 595/2009 as regards the certification of the CO2 emissions and fuel consumption of heavy-duty vehicles and amending Directive 2007/46/EC of the European Parliament and of the Council

An Experimental Study of the Underbody Flow of a VW Golf VII Under On-Road and Wind-Tunnel Conditions

Johannes Haff[1(✉)], Sven Lange[1], Tarik Barth[1], and Henning Wilhelmi[2]

[1] Development Aerodynamics, Volkswagen AG, Wolfsburg, Germany
Johannes.Haff@volkswagen.de
[2] Institute of Aerodynamics and Flow Technology, German Aerospace Center, Göttingen, Germany

1 Introduction

The further reduction of the aerodynamic drag of today's passenger cars requires the development of regions such as the vehicle underbody, the wheel housings and the tires. Further, an understanding of the interference effects of the engine cooling flow with those flow regions can help to improve the vehicle flow to reduce drag.

1.1 Objective of This Study

This study is aimed at investigating the influence of the engine cooling flow on the flow field underneath a VW Golf VII. Another goal is to identify the effect of drag reducing devices such as a radiator shutter and a tunnel paneling on the ground flow physics.

In addition to wind tunnel experiments and CFD studies, a road test was performed to compare the velocity field underneath the vehicle to that induced under the on-road conditions on a test track.

1.2 Procedure

The velocity field underneath a VW Golf VII was measured using the laser-based optical measurement method particle image velocimetry (PIV). The test vehicle was therefore equipped with an on-board camera and laser system.

Road tests were performed on a test track of the test track facility Ehra-Lessien of the Volkswagen Group. The same instrumented test vehicle as used in the road test was also used in the wind tunnel experiments, which were performed in the Vehicle Aeroacoustics Wind Tunnel of the IVK Stuttgart. The CFD simulations were based on the Lattice-Boltzmann solver Powerflow from Exa.

J. Wiedemann (ed.), *Progress in Vehicle Aerodynamics and Thermal Management*,
https://doi.org/10.1007/978-3-319-67822-1_12

2 Experimental Setup

2.1 Test Vehicle

A VW Golf VII served as test vehicle for the experimental study. The vehicle dimensions are 4.3 m in length, 1.5 m in height and 2 m in width. The car was equipped with 15″ steel wheel rims and 195/65 R15 Continental ContiEcoContact 3 tires. Based on the vehicle length, the standard Reynolds number was 10.7 million, which corresponds to a driving velocity of 140 km/h. The Reynolds number was varied in a range of $6.2 \cdot 10^6 \leq Re \leq 12.3 \cdot 10^6$.

To set the specific ground clearance for a standard chassis of a VW Golf VII, a stiffer suspension system was required due to the additional weight of the measurements and supply systems.

2.2 Vehicle Configurations

In addition to the baseline vehicle configuration, a radiator shutter was simulated by closing the lower cooling inlet at the vehicle front (see Fig. 1, left). To quantify the effect of the tunnel paneling on the velocity field, for the third vehicle configuration the tunnel paneling, framed in green in Fig. 1, right, was dismounted.

Fig. 1. Vehicle configuration with closed lower cooling air inlet (left) and view on the vehicle underbody with the tunnel paneling framed in green (right)

2.3 Vehicle Instrumentation

As described in the further course of this paper, the PIV system was installed as an on-board system in the vehicle. Compared to a track-fixed PIV system this offers the opportunity to generate more measurements results during a single test run for data averaging purposes. Further, with a stationary setup the correct position of the light sheet plane relative to the vehicle's underbody cannot be guaranteed.

On the other hand, a vehicle-fixed setup involves a higher demand on the experimentalist. Contaminating effects such as the relative motion between the camera system and the vehicle as well as the displacement of the laser light sheet during testing have to be considered in the data analysis.

In addition to the camera and laser systems of the PIV setup, the vehicle was also equipped with pressure taps located at the vehicle front and the ground clearance was measured during the road test using optical sensors installed on the vehicle underbody. The following section (Sect. 2.4) discusses the measurement setup in detail.

The power supply and the data acquisition were also a part of the interior equipment. Two 100 Ah batteries served as energy source during testing. A current transducer was used to convert the 12V-DC voltage signal into a 230V-AC voltage signal to power the PIV and data acquisition systems.

2.4 Measurement Techniques

2.4.1 Particle Image Velocimetry

2.4.1.1 Measurement Principle. Particle image velocimetry is a laser-based optical measurement method to determine the flow velocity from a measured particle displacement at a specific time delay. Therefore, seeding particles are injected into the investigated flow region. A pulsed laser in conjunction with an optical lens system is used to illuminate a distinct flow region at two different times. The laser light is scattered by the induced seeding particles and a digital camera records two separate particle images. Cross-Correlation algorithms are applied to calculate the particle displacement. In conjunction with the known time delay between the two image recordings, the resulting local flow velocity can be calculated. Detailed information about PIV can be taken from [1].

2.4.1.2 Measurement Setup. In this study, a 2-component PIV setup was used to measure the two in-plane velocity components in the vehicle's symmetry plane underneath the test vehicle. Two CCD cameras type PCO2000 with an area resolution of 2048 × 2048 pixels and a 14 bit dynamic range were installed on a mounting device on the co-driver's side door as can be seen in Fig. 2, left. The cameras were aligned in driving direction to reduce flow contamination induced by the camera systems due to blockage.

Fig. 2. Instrumented test vehicle with a mounting device for the two CCD cameras installed on the vehicle's co-driver side (left) and laser light sheet in the symmetry plane of the vehicle's underbody (right)

Mirrors were used to deflect the view direction by 90° perpendicular to the driving direction to measure the longitudinal and vertical flow velocity in a vertical streamwise cut below the test vehicle (see Fig. 3). The camera power supplies were placed inside the vehicle and connected via cable connections to the CCD cameras. With a 35-mm camera lens, the camera field of view was 0.5 × 0.19 m^2 and the total view of both cameras 0.98 × 0.19 m^2 with an overlap of 20 mm in the streamwise direction (see Fig. 3). With an interrogation window size of 24 × 24 pixels, the resultant resolution of the velocity vectors was 1.7 mm.

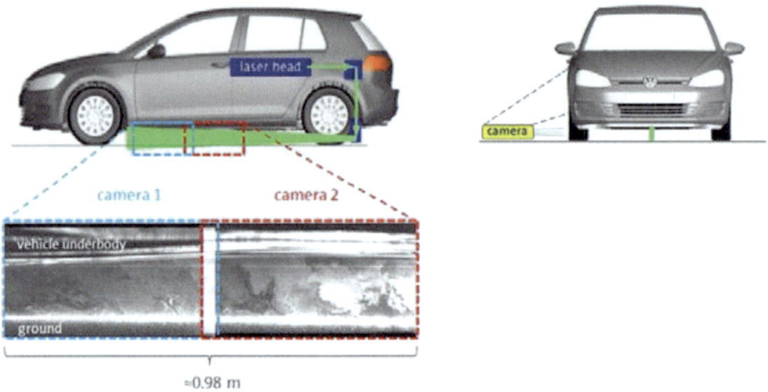

Fig. 3. Schematic sketch of the PIV setup with the camera field of view and the position of the laser light sheet relative to the test vehicle

In addition to the camera system, a major challenge was the illumination of the measurement plane in the vehicle's underbody during road testing. An air-cooled double-pulse Nd:YAG laser system with a pulse energy of 200 mJ from Litron Lasers was used. The laser system was installed inside the vehicle. The limited space required an installation of the laser head parallel to the optics bar. Two coated mirrors deflected the laser beam by 180°. A spherical lens system behind was used to adjust the laser beam thickness to 2 mm in the measurement region. With a third mirror the laser beam was deflected downwards through a metal tube outside the vehicle and a cylindrical lens inside the tube generated a light sheet plane. A forth mirror at the end of the tube deflected the laser light sheet in a vertical streamwise plane with regard to the vehicle's coordinate system. Laser and camera systems were triggered by a sequencer PC.

For laser safety reasons, a timberwood plate was installed between the front and rear part of the greenhouse to shield the greenhouse front from the laser light.

Dual-frame images with a sampling frequency of 15 Hz were recorded by the camera systems. A trigger signal controlled the camera and the laser system. Each measurement was manually triggered either after reaching constant flow conditions in the wind tunnel or after reaching a constant driving velocity in the road test.

2.4.2 Pressure Measurement System

To measure the atmospheric conditions during the road test a stationary measurement unit was installed close to the test track. Since the results of this track-fixed unit gives no insight into the local inflow conditions of the test vehicle during road testing, the test vehicle was equipped with static pressure taps at the front bumper and in the vehicle's symmetry plane of the engine bonnet and the roof in order to detect cross-wind effects.

A digital temperature compensated (DTC) electronic pressure scanning system type ESP-64HD from Esterline Technologies Corporation recorded the pressure signals. The sampling frequency was set to 100 Hz. As reference pressure, a pressure tap on the engine bonnet were the pressure reaches atmospheric conditions was used.

The pressure results were used as criterion for the PIV results. Only velocity data at no wind conditions were considered in the data averaging.

2.4.3 Ground Clearance Measurement System

Another important aspect of the road test is the consideration of the ground clearance. Optical sensors based on the triangulation principle were installed close to each wheel to measure the local ground clearance during the road test. By the use of a stiffer suspension system and additional weights inside the vehicle, the ground clearance was set to the standard chassis height.

In addition to measurements by the use of optical sensors, the results of PIV was also used to measure the mean ground clearance, as it will be discussed in the further course of this article.

2.5 Road Test

2.5.1 Test Track

Measurements were performed on a test track on the vehicle dynamics test area of the test track facility Ehra-Lessien of the VW Group (Fig. 4). The closed-loop test track is characterised by two 180° curvatures connected by two straight parts of around 1 km in length.

PIV was performed over a length of around 200 m on the western of one the straight parts of the test track.

Fig. 4. Vehicle dynamics test area on the test track facility Ehra-Lessien of the Volkswagen Group

2.5.2 Stationary Measurement Systems

After the presentation of the vehicle instrumentation, the application of track-fixed measuring systems is discussed in this section. For PIV, a flow seeding is required for illumination purposes. Since an on-board seeding system may disturb the vehicle flow, it was decided to use a stationary system on the test track. A large-volume Laskin nozzle aerosol generator powered by a diesel compressor atomised sunflower oil particles of a mean diameter of 0.4 to 0.5 μm. With pipes located at different positions close to the test track an homogeneous seeding distribution was generated along the test track.

To consider the effect of changing environmental conditions during testing, the ambient conditions in terms of pressure, temperature and humidity were measured at different positions of the test track. A hydrometric vane and a wind vane located at halfway the PIV test track provided the occurring wind conditions.

2.5.3 Test Procedure

The track-fixed flow seeding was switched on and the test vehicle was accelerated on the southern curve of the testing area. After reaching a constant driving speed, the camera and laser systems were activated before reaching the PIV test track. Measurements were performed at a sampling frequency of 15 Hz of both, the camera and the laser system. The co-driver controlled the measurement and data acquisition systems.

All measurements were performed during darkness to avoid a reduction in contrast of the particle images due to daylight. Side lights were used to illuminate the test track.

2.6 Wind Tunnel Experiment

2.6.1 Test Facility

The wind tunnel measurements were performed in the Vehicle Aeroacoustics Wind Tunnel of the IVK Stuttgart. The closed-circuit, single-return, subsonic wind tunnel with an open test section is driven by a 3.3 MW engine located in the return part of the tunnel. In the test section with a nozzle exit plane of 5.80 m × 3.87 m a maximum flow velocity of 260 km/h can be reached. The facility is equipped with a ground simulation system consisting of a boundary layer pre-suction and a 5-belt system.

2.6.2 Wind Tunnel Setup

The same instrumented test vehicle as used for the road test was also used in the wind tunnel experiment. Contrary to the road test, the energy supply of the measuring systems was realised from outside the test section. Therefore, a cable connection was located on the right side of the test vehicle as can be seen in Fig. 5, left. The test vehicle was fixed with fastening sleeves close to the wheels.

Two Laskin nozzle aerosol generators were located in the settling chamber of the wind tunnel to generate the flow seeding. Particles of DEHS (Di-Ethyl-Hexyl-Sebakat) with a mean diameter of 0.4 to 0.5 mm were injected into the flow. By running the tunnel over a time period if several minutes a homogeneous seeding density was reach in the complete wind tunnel.

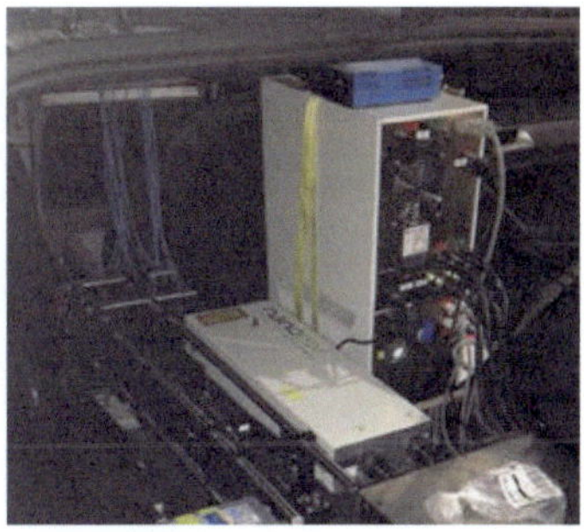

Fig. 5. Test vehicle in the test section of the Vehicle Aeroacoustics Wind Tunnel of the IVK (left) and view inside the test vehicle with the battery pack at the bottom left and the PIV system at the top (right)

3 Experimental Results

3.1 Results of the Wind Tunnel Experiments

As part of the analysis process, the influence of the number of single measurements on the average was investigated. The discussed results are based on a data average with a maximum deviation of less than 2%.

Figure 6 visualises the vertical flow velocity as coloured contours superimposed by a mean velocity streamline plot. The tunnel paneling and the simulated radiator shutter revealed the expected influence on the vehicle's underbody flow. The tunnel paneling forces the engine cooling flow to leave the engine compartment parallel to the free-stream direction and therefore reduces pulse losses. When the paneling is removed, the direction of the vertical velocity component is changed. As can be seen in Fig. 6, the underbody flow of the baseline case (top) characterised by an upward directed vertical velocity component is deflected downwards when the tunnel paneling is removed (bottom).

The cooling flow therefore shields the vehicle underbody's flow, which reduces the mass flow rate through the underbody region in this section cut. In the velocity profiles in Fig. 7, the reduced streamwise velocity component for the configuration without the tunnel paneling is clearly visible. The x-position is based on the vehicle's coordinate system. The position X = 1 m corresponds to a distance of 0.16 m behind the tunnel paneling.

The simulated radiator shutter has the opposite effect. The streamwise flow velocities increase as visualised in Fig. 7.

3.2 Comparison of Wind Tunnel Experiment, Road Test and CFD

The results of the wind tunnel experiment and the road test are in good agreement to each other. The comparison of the baseline case is presented in Fig. 8 as coloured contours of the streamwise velocity component (U) and the mean flow direction visualised as velocity vectors. It has to be considered that only every third velocity

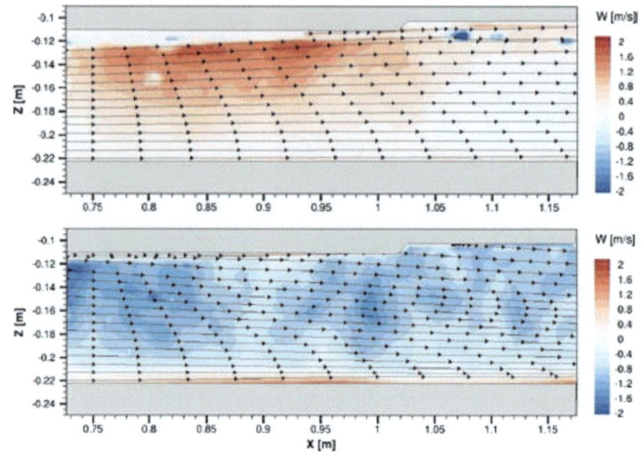

Fig. 6. Vertical velocity contour plot superimposed by a mean streamline plot measured in the wind tunnel experiment for the baseline configuration (top) and for the configuration without tunnel paneling (bottom). The averaged results are based on 700 single measurements of the unsteady flow field

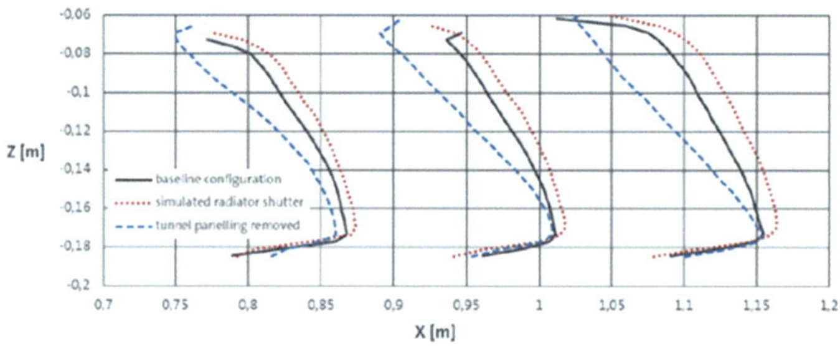

Fig. 7. Velocity profiles of the baseline configuration (black), the configuration with a simulated radiator shutter (red) and the configuration without tunnel paneling (blue) measured at X = 1 m in the wind tunnel experiment.

vector is presented in the visualisations. The results agree both from the qualitative and quantitative point of view.

Figure 9 visualises velocity profiles at different longitudinal positions in the underbody region. The same vertical distribution of the streamwise velocities was measured in the wind tunnel experiment and the road test. Further, the absolute values also agree.

Since the main objective of this article is the discussion of the results provide by experiments in an automotive wind tunnel and on a test track, the setup of performed CFD studies is not discussed in deep detail. For the sake of completeness, some results are presented in the following. The simulations were performed using the Lattice-Boltzmann flow solver Powerflow from Exa. A large domain representing the on road conditions was used. The rim rotation was simulated by the use of a sliding

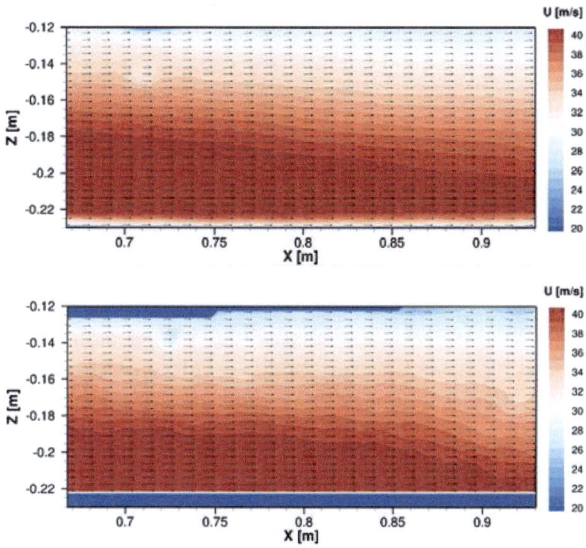

Fig. 8. Contour plot of the streamwise velocity (U) superimposed by the every third vector of the mean velocity vector field measured in the wind tunnel experiment at FKFS (top) and in the road test (bottom)

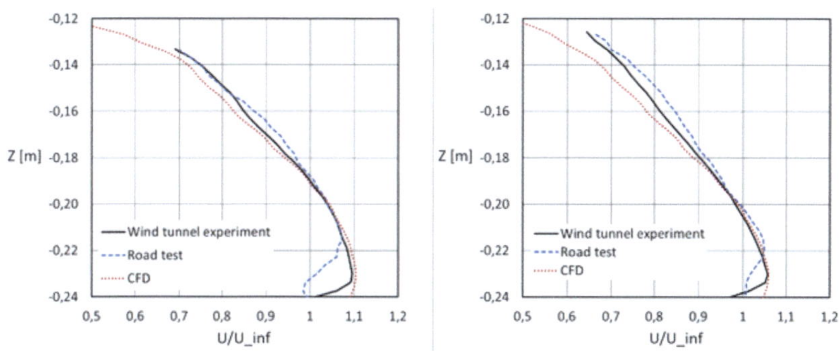

Fig. 9. Velocity profiles of the baseline configuration measured in the wind tunnel experiment (black), the road test (blue) and predicted by CFD (red) at a streamwise position of X = 0.75 m (left) and X = 0.9 (right)

mesh. The relative motion of the tyre and the ground with regard to the vehicle geometry was set with boundary conditions. The test vehicle was laser-scanned to use the same geometry in the simulation as used in the experiments. The camera mounting on the co-driver's side and the laser optics on the vehicle underbody were therefore also considered in the simulation.

4 Conclusions

The velocity field was measured using the laser-based optical measurement method particle image velocimetry (PIV) in a wind tunnel experiment and a road test. The results were used to investigate the influence of a tunnel paneling and a simulated radiator shutter on the flow characteristics underneath a VW Golf VII. Further, the capability of a modern automotive wind tunnel equipped with a state-of-the-art ground simulation system was verified by using the results measured in a road test on a test track.

The velocity fields measured in the wind tunnel experiments reveal a distinct influence of the investigated vehicle configurations on the underflow aerodynamics. A significant effect on the mass flow rate could be detected. The analysis of the wind tunnel results further provides a good correlation to the flow velocities measured in the road test.

The comparison to flow field predictions from numerical simulations reveal a higher velocity gradient in the horizontal and the vertical direction.

Reference

1. Raffel, M., Willert, C., Kompenhans, J.: Particle Image Velocimetry – A practical Guide. Springer (1997)

Some Aspects on On-Road Aerodynamics

Thomas Schütz(✉) and Hannes Vollmer

BMW AG, Knorrstraße 147, 80788 Munich, Germany
{Thomas.ST.Schuetz,Hannes.M.Vollmer}@bmw.de

Abstract. The aerodynamics of modern day passenger cars are usually developed under constant and reproducible conditions in a wind tunnel. However, spontaneously occurring wind changes during on-road drive are of significant interest. Recently, BMW AG has been able to develop virtual methods enabling the investigation of such spontaneous on-road driving events. These methods have been used to assess driving in slipstream, cornering and the impact of naturally occurring wind gusts. During slipstream driving, a noticeable drag-reducing effect was found for a passenger vehicle behind a truck and behind different sized passenger cars. There was a significant effect on vehicle lift. Cornering conditions were examined by only changing the curvature of the streamlines in the flow field. Simulation results indicate a significant change on vehicle coefficients compared to straight line driving. Lastly, the flow in gusty wind conditions was investigated. Simulation results confirmed results from academic wind tunnel measurements. An increase in the aerodynamic yaw moment was found under transient oncoming flow conditions compared to the quasi-steady inflow. Overall, the results from the three investigated driving conditions have a significant relevance of the aerodynamic effects under real driving conditions.

1 Introduction

The most important part of the aerodynamic development of passenger cars is the optimization of the vehicle drag, which has been discussed in depth, e.g. [6, 9, 10]. However, there are further development tasks which must be solved within the scope of aerodynamic vehicle development. The driving characteristics are especially influenced by low values for the aerodynamic lift, the side force and the yaw moment. According to BMW's Efficient-Dynamics strategy, the gain of knowledge on all these parameters plays an important role to developing a vehicle with good handling.

In order to measure the air drag to determine either the legal or the customer's average fuel consumption, it is sufficient to consider time averaged values. Short-term deviations from a standardized quasi-steady oncoming flow are not significant. However, this can be quite relevant for the prognosis of a realistic individual instantaneous consumption. Furthermore, the safety-oriented development of the driving dynamics requires an examination of short-term wind load events. For this purpose, aerodynamics should also be understood and optimized for all situations occurring during on-road driving.

Figure 1 shows an overview of the driving situations assessed in this study apart from standard driving conditions. The first situations examined are slipstream

J. Wiedemann (ed.), *Progress in Vehicle Aerodynamics and Thermal Management*,
https://doi.org/10.1007/978-3-319-67822-1_13

conditions that occur during the column drive. Results from the literature are already known for the pursuit of a truck and the arrangement of vehicle-like basic bodies arranged one behind the other and are summarized in [10]. In the current study, all vehicle distances relevant for real driving operation were considered. The slipstream effect behind different types of cars of various sizes was also examined besides slipstreaming behind a truck.

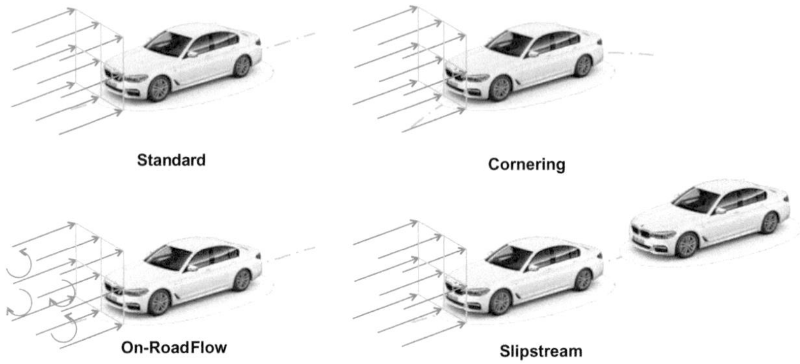

Fig. 1. Considered on-road driving situations

From an aerodynamic view the issue of cornering consists of several parts. The streamlines observed by the vehicle are curved and is an important change to the flow field. The change of the vehicle body relative to the road due to the occurrence of the rolling moment as well as the adjustment of the chassis due to the occurring steering angle also affect the aerodynamics. However, both are neglected within the scope of the present study.

The influence of natural wind gusts on aerodynamic vehicle characteristics has recently been reported in [11, 12]. In the case of truck aerodynamics, the temporal weighting of occurring wind events by their occurrence probability is a state of the art development. The c_D value for a given vehicle is weighted by the probability and the length of time that a given wind event will occur. In the context of this study, similar probabilities of natural wind speeds near the ground were analyzed for different regions in Germany on the basis of available meteorological data. From this, a time-averaged lateral wind inflow could be evaluated for different vehicle types. Furthermore, the influence of a gusty oncoming flow was assessed by computational fluid dynamics (CFD) and the significance for various aerodynamic parameters in the frequency domain was evaluated. This procedure is known from the latest experimental investigations published by the University of Stuttgart, [11]. The following section discusses the methods that were used for these investigations.

2 CFD Method

The investigation of slipstream was carried out using the BMW CFD method standard with the software package EXA PowerFlow, which has been described in many publications, e.g. [2, 3, 5, 8]. However, the current BMW best-practice setup applies only to standard flow conditions. Placing two vehicles at the desired distance from each other can thus easily represent the slipstream situation.

For the purpose of cornering, the vehicle is positioned with its coordinate origin in the axis of a cylinder with the radius R. The cylinder is defined as non-inertial reference system, whereby, the movements of the internal geometries can be simulated. The mapping of cornering is achieved by a rotation of this reference system, where $\omega \cdot R = V_F$ meaning that the center of the curve lies at the edge of the cylinder with the radius R, cf. Fig. 2.

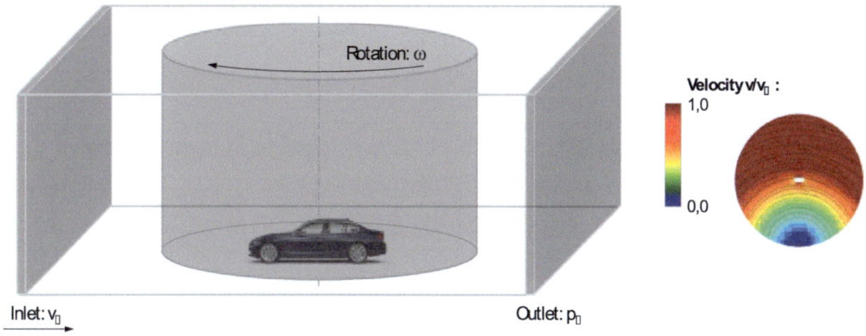

Fig. 2. Procedure of the CFD setup for cornering investigations

For on-road flow conditions, the turbulent structures are applied at the flow inlet to simulate gusty oncoming flow. These are based on on-road measurements of Wordley and Saunders [12] and the PowerFLOW® implementation by Duncan et al. [1]. Therefore a constant ratio of the turbulent intensity and length dimension components is used ($I_x = I_y = 7{,}35 \cdot I_z$ as well as $L_x = L_y = 1{,}3 \cdot L_z$). Thus, only the x components are used here to describe the input parameters of a gust simulation, i.e. $I = I_x = 7\%$, $L = L_x = 2{,}5\,\mathrm{m}$.

In contrast to this gust setup, the BMW standard setup places the flow inlet 100 m upstream of the vehicle. In order to compensate for the effect of shortening the simulation volume, the speed and pressure values of a quasi-steady computation with the BMW standard setup are applied to each voxel in addition to the turbulent structures mentioned above. This calibration ensures that there is a match between the results of a gust simulation with $I_x = I_y = I_z = 0\%$ and $L_x = L_y = L_z = 0$ and quasi-steady simulation with the BMW standard setup.

3 Analysis of Slipstream

For all the evaluated slipstream combinations, a BMW 3 Series sedan (F30) was used as the pursuing vehicle for all simulations. The FAT truck, the identical BMW 3 Series Sedan (F30), a BMW X6 (F16) and a Mini hatch back (F56) were used as the leading vehicles. The frontal area ratios of the cars as well as the vehicle distances examined for the analysis are shown in Fig. 3. The law prohibits very small distances for tailgating another vehicle. A separation distance up to 140 m was considered when using the leading BMW F30, 80 m for the FAT truck and up to 40 m for the other vehicles.

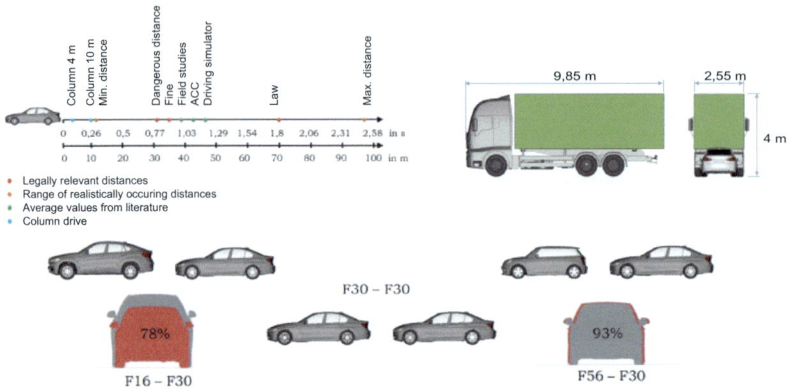

Fig. 3. Overview on investigated distances and vehicle combinations

The effect on the leading vehicle is irrelevant for legally permissible separation distances [7]. In the case of a lateral offset, there are similar values at legally permissible distances as for the pursuit condition. Detailed results are given by Huber [3].

Figure 4 shows the influence of the distance and the vehicle configuration on the aerodynamic parameters drag, front axle and rear axle lift. Regardless of the vehicle combination, the drag of the pursuing car is significantly reduced. This effect is particularly evident if the truck is being followed.

This is due to the dead water area and the resulting total pressure drop. The drag reduction for F56 and F30 also scales with the vehicle size while this trend is not seen with the F16. However, this can be explained from the flow topology shown in Fig. 5, where the coupé-shaped rear end of the F16 reduces the wake. Thus, the drag reduction in the slipstream is also smaller. These results indicate that overall size of the dead water is a good scaling variable for the estimating the drag reduction. Overall, drag improvements of 20 to 50 counts are possible when the pursing vehicle is 20 m to 40 m behind. The effects of the slipstream drive on the lift of the pursuing vehicle are noticeable. The effect is similar for all vehicle combinations which a decrease in front lift and an increase in rear lift of between 20 and 40 counts, which causes a severe pitching moment. This observation can also be explained by the wake structure in Fig. 5, since the front of the pursuing vehicle is subjected to a downwash flow.

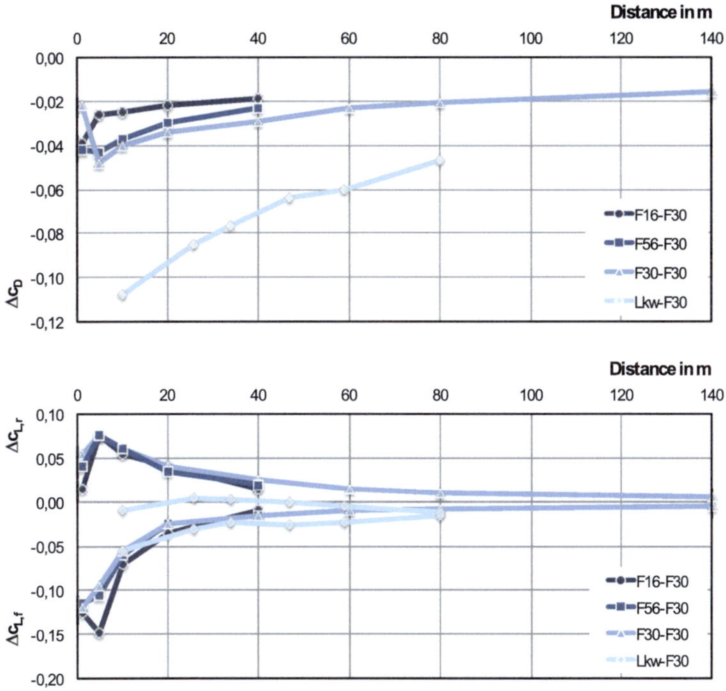

Fig. 4. Drag coefficient (upper), front and rear axle lift coefficient (bottom) vs. vehicle distance for the four pursuing vehicles

This leads above all to a significant reduction of the front axle lift. This is most pronounced when a coupe-shaped rear end (F16) vehicle is in the lead.

4 Effects of Cornering

To assess the cornering influence on aerodynamics, three corner radii were investigated. Table 1 shows the values chosen and the corresponding curvature, defined as the reciprocal value of the radius.

The impact of cornering on the aerodynamic coefficients of drag, front and rear axle lift are shown in Fig. 6. All three coefficients increase with increasing curvature. Even with a radius of 20 m, the drag coefficient increases by about five counts and the front axle lift coefficient increases by 50 counts. This can lead to a noticeable influence on driving dynamics and fuel consumption. The change in rear axle lift can is rather small in comparison.

The difference of the aerodynamic characteristics of the vehicle can be seen in the flow topology in Fig. 7 when comparing straight and corner conditions. Both the iso-surfaces, as well as the streamlines in the wake structure show an asymmetrically distorted flow field.

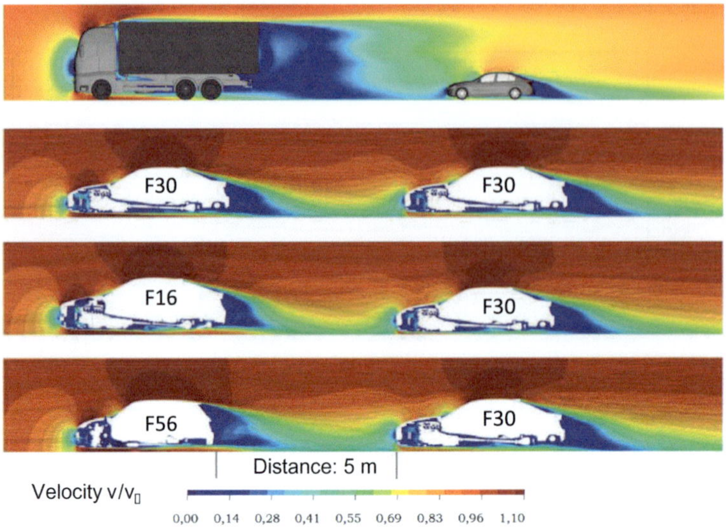

Fig. 5. Flow topology for slipstream conditions, represented by the distribution of the flow velocity in the vehicle center section

Table 1. Parameters of the investigated corners

Radius in m	∞	20	10	5
Curvature in m^{-1}	0,00	0,05	0,10	0,20

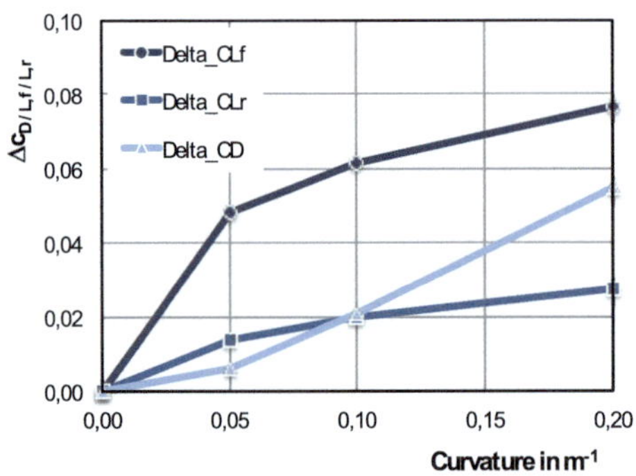

Fig. 6. Air drag and axle lift coefficient depending on the corner curvature

5 Natural Wind Effects

Driving in gusty oncoming flow has been investigated using a probabilistic-based computation method as well as numerical flow simulations. Kopp and Frank [4], provided a classical approach to the assessment of natural wind on vehicle

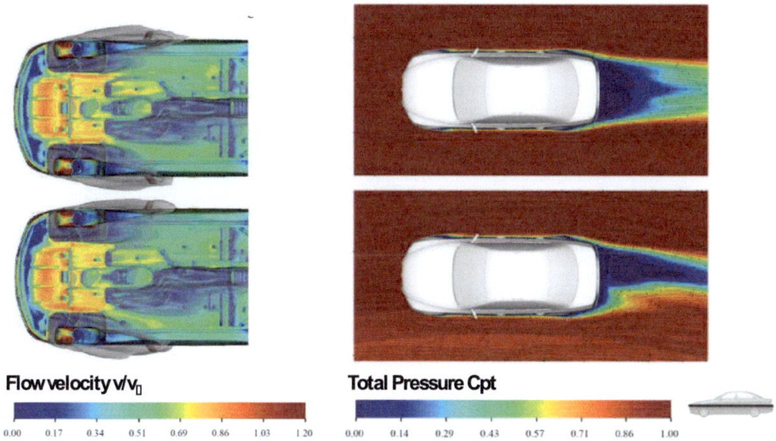

Fig. 7. Flow topology during cornering (bottom), represented by the near-wall flow velocity and the total pressure iso-surfaces (left) as well as the distribution of the total pressure in a plane z = const.

aerodynamics. For this purpose, the yaw angle occurring during driving is included as a probability function of natural wind speeds. Figure 8 shows the geometric conditions for determining the yaw angle. Also shown is the probability distribution of wind velocities for three German cities with different geographic conditions. These are approximated by Weibull distributions with weather data from the DWD, [13].

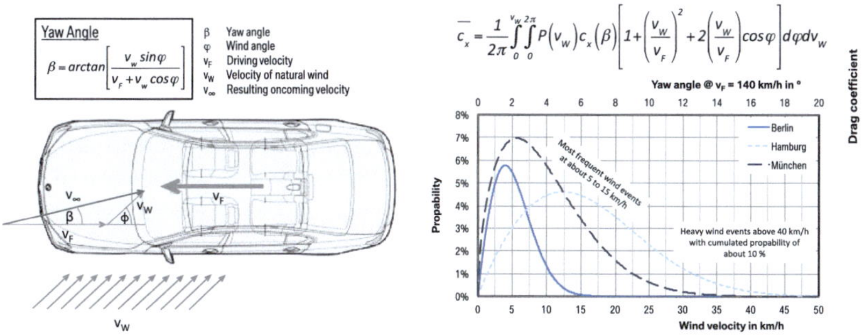

Fig. 8. Definition of some side wind-related flow variables (left), frequency distribution of wind events in Germany (right), evaluated by data from [13]

The equation in the upper right of Fig. 8 shows how a probability-weighted drag coefficient can be determined with this data. In addition to the probability distribution P (v_W), it is assumed that the wind direction ϕ is distributed equally in all directions during the sum of all rides. Furthermore, the yaw angle dependency of the drag coefficient is also required as an input variable in the equation, cf. Fig. 9.

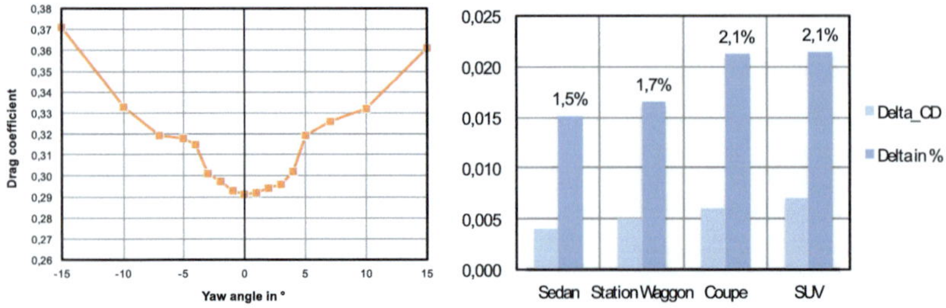

Fig. 9. Example of the yaw moment distribution of the drag coefficient (left), wind occurrence propability-weighted drag coefficients for different vehicles (right)

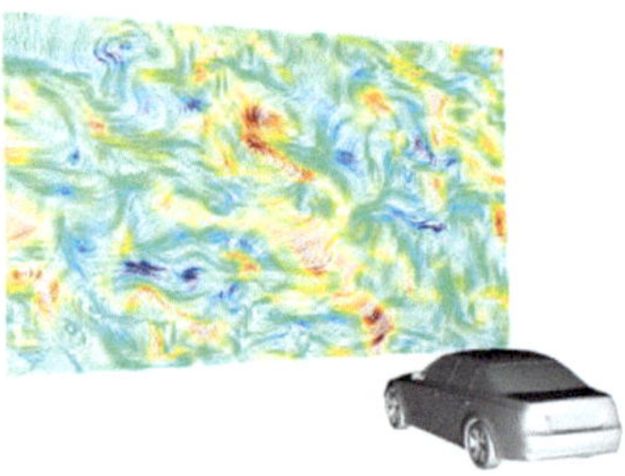

(a) $I = 10\% = \text{const.}, L = 2{,}5\,m.$

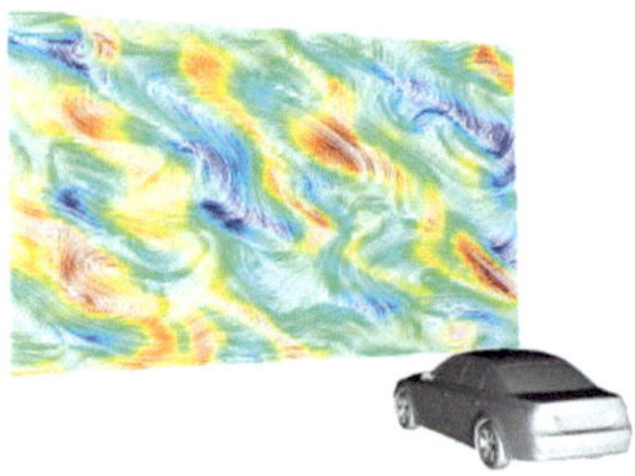

(b) $I = 10\% = \text{const.}, L = 5\,m.$

Fig. 10. Visualization of the gusty flow with the CFD method introduced in [1]

Using all the mentioned input parameters, weighted drag coefficients are shown in Fig. 9 (right). For all the different vehicle geometries, there is an increase of between four and seven counts (1.5 to 2.1%) compared to cases with a uniform flow field. Since, the occurrence probability of the wind events has been included in this consideration, they can be directly input into the consumption calculation (in contrast to results from the cornering and slipstream investigations).

A more elaborate assessment procedure for the consideration of natural wind conditions was presented by Duncan et al. in [1]. According to Wordley and Saunders [12], Fig. 10a represents a realistic on-road drive. The effects of the realistic flow on the aerodynamic force coefficients as a difference to the quasi-steady baseline case are shown in Fig. 11.

Of note is the massive increase in drag and front axle lift as a function of the degree of turbulence. The rear axle lift is significantly increased only at values of low turbulence. Also shown is that the turbulence length scale is less critical than Degree of Turbulence on the aerodynamic coefficients.

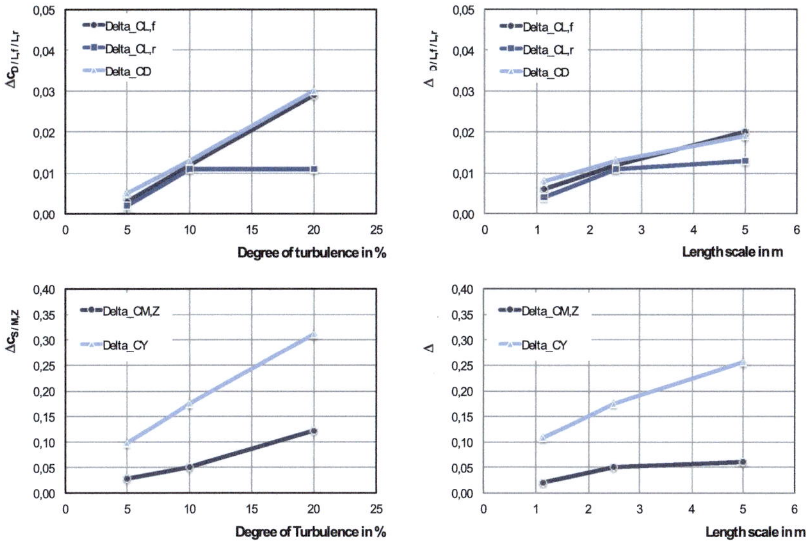

Fig. 11. Time-averaged drag and lift coefficients as well as maximum values of the yaw moment and the side force for gusty oncoming flow, compared with the quasi-steady reference case

Driving dynamics also depend on temporal variation of the yaw moment and the side force. In the case of zero degree averaged yaw angle, only the load peaks occurring over time are considered. In the case of gusty airflow, the side force increases significantly, especially as the degree of turbulence increases (300 counts). Side force was more sensitive to turbulence length scale, with increases up to 250 counts. Consequently, a less sovereign smooth and stable handling behavior will be felt by the driver.

Acknowledgement. Many thanks go to Elisabeth Huber [3], Markus Geisenhofer [2] and Jose Pabon Andrade [5], who helped to investigate the above mentioned aerodynamic phenomena during their time as master and bachelor students at BMW.

References

1. Duncan, B., Gargoloff, J., Alajbegovic, A.: On-road aerodynamic efficiency under realistic transient wind conditions. Exa Corporation White-Paper, Burlington MA, USA (2014)
2. Geisenhofer, M.: Aufbau einer Simulationsmethode zur Bewertung von instationärer Aerodynamik. Master-Arbeit, TU Darmstadt (2015)
3. Huber, E.: Untersuchungen zum Interferenzwiderstand bei Kolonnenfahrt von Kraftfahrzeugen. Bachelor-Arbeit, TH Ingolstadt (2015)
4. Kopp, S., Frank, T.: Nutzfahrzeuge. In: Schütz, T. (Hrsg.) Hucho – Aerodynamik des Automobils. 6. Auflage. Springer Vieweg, Wiesbaden (2013). ISBN: 978-3-8348-2316-8
5. Pabon Andrade, J.: Aerodynamische Analyse von Pkw bei Kurvenfahrt. Bachelor-Arbeit, TU München (2016)

6. Schütz, T.: Fahrzeugaerodynamik – Basiswissen für das Studium. 1. Auflage. Springer Vieweg, Wiesbaden (2016). ISBN: 978-3-658-12817-3
7. Schütz, T.: WLTP – On the increased importance of aerodynamics and impact on development procedures. In: Documentation Volume 1. 16th Stuttgart International Symposium, Stuttgart (2016)
8. Schütz, T., Grün, N., Blumrich, R.: Numerische Methoden. In: Schütz, T. (Hrsg.) Hucho - Aerodynamik des Automobils. 6. Auflage. Springer Vieweg, Wiesbaden (2013). ISBN: 978-3-8348-2316-8
9. Schütz, T., Klußmann, R., Neuendorf, R.: Automobile Aerodynamik 2020. ATZ - Automobiltechnische Zeitschrift Ausgabe Nr.: 2016-12
10. Schütz, T., Krüger, L., Lentzen, M.: Luftkräfte und deren Beeinflussung an Personenkraftwagen. In: Schütz, T. (Hrsg.) Hucho – Aerodynamik des Automobils. 6. Auflage. Springer Vieweg, Wiesbaden (2013). ISBN: 978-3-8348-2316-8
11. Stoll, D., Kuthada, T., Wiedemann, J., Schütz, T.: Unsteady aerodynamic vehicle properties of the DriveAer model in the IVK model scale wind tunnel. In: 10th FKFS Conference, Stuttgart (2015)
12. Wordley, S., Saunders, J.: On-road Turbulence: Part 2. SAE Technical Paper Series, No. 2009-01-0002 (2009)
13. ftp://ftp-cdc.dwd.de/pub/CDC/observations_germany/climate/hourly/wind/recent/ aufgerufen am 22. Mai 2016

Aerodynamic Development

On the Influence of Underhood Flow on External Aerodynamics of the DrivAer Model

Christopher Collin[1(✉)], Jörg Müller[1], Moni Islam[1], and Thomas Indinger[2]

[1] AUDI AG, 85045 Ingolstadt, Germany
christopher.collin@audi.de
[2] Technical University of Munich, Munich, Germany

Abstract. The interaction of underhood flow with external aerodynamics is part of the main topics for the development of automotive aerodynamics. In particular, the location and design of cooling air exits as well as the mass flow through the radiator package are considered to be important factors. The DrivAer is an open-source, aerodynamic reference model, which has a realistic overall geometry compared to rather simple, generic shapes such as the SAE body. The model is flexible in use as it provides different rear end and wheel designs. The open-grille upgrade of the DrivAer enhances the potential of the body towards a reference model for more detailed, open cooling investigations.

In the present study, the 40% scale Mock-Up DrivAer model of the Technical University of Munich was upgraded with underhood flow for wind tunnel tests with moving ground simulation. The wind tunnel measurements covered different underhood flow configurations including variations for the pressure drop of the radiator package and different locations for cooling air exits – both for different rear end shapes of the model. Additionally, the effect of underhood flow on the aerodynamics of both the notchback and the estateback rear end were calculated for six different radiator packages using DDES simulations with OpenFOAM®.

1 Introduction

The open-source DrivAer reference model [1] has been subject of a variety of investigations with respect to the development of automotive aerodynamics. The DrivAer geometry was introduced by Heft et al. [2]. After the upgrade of the large wind tunnel of the Technical University of Munich (TUM) with moving ground simulation several experimental investigations on DrivAer aerodynamics have been performed using the 40% scaled model, e.g. [3–9]. Mack et al. [4] firstly presented an interior geometry for underhood flow investigations and Heft et al. [5, 7] showed both experimental and numerical results for a variation of cooling inlets and outlets.

Wittmeier and Kuthada [10] introduced a new, more realistic, version of an open cooling DrivAer body, which was firstly upgraded with a defined engine compartment geometry. The geometry is open-source and available for download at TUM [1]. Wittmeier and Kuthada [10] presented first results for the 25% scale open grille model.

J. Wiedemann (ed.), *Progress in Vehicle Aerodynamics and Thermal Management*,
https://doi.org/10.1007/978-3-319-67822-1_14

Furthermore, both numerical and experimental results for the DrivAer with underhood flow are presented by Kuthada et al. [11], Matsumoto et al. [12], Simmonds et al. [13] and Haag et al. [14].

In the present study, the 40% scale open cooling geometry was selected in order to achieve a better understanding about the effects of underhood flow on the aerodynamics of the DrivAer. Therefore, both experimental and numerical methods are applied. The underhood flow conditions were adapted using the following parameters:

- Pressure drop characteristics of the radiator package
- Locations of the underhood flow outlets
- Rear end shape of the body
- With/without moving ground simulation

In preparation of the experimental investigations the existing 40% scale DrivAer model was upgraded with the new underhood flow compartment. Therefore, a modular package for the radiator was installed in order to test different configurations for the pressure drop of the module.

Unsteady simulations with DDES were performed using OpenFOAM® for some selected configurations from the measurements. A comparison of results from both the experimental and the numerical approach are presented.

2 Experimental Approach

The aerodynamic tests were conducted in the Wind Tunnel A (WTA) – the large wind tunnel of the TUM. In the following, the experimental setup of the wind tunnel and the DrivAer model as well as the experimental results are presented.

2.1 The 40% Scaled DrivAer Model in Underhood Flow Configuration

In the first step of the present study the 40% scale DrivAer model of the Technical University of Munich (TUM) was updated with the underhood flow geometry. This upgrade included the integration of an adapted underfloor inset into the existing model. Figure 1 depicts the main differences between the model in Mock-up and underhood flow configuration on the outer shell of the body. As presented by Wittmeier and Kuthada [10], a lower and an upper open grill serve as inlets for the underhood flow, whereas the outlets are located inside the front wheel houses and in the transmission tunnel.

The radiator package was designed as a stack in order to easily adjust the pressure drop, see Fig. 2. The main components such as the inlet duct, housings for the flow straighteners, perforated sheets and fan shroud are milling parts. The connection of each component is realized with sealing frames made of synthetic EPDM rubber. Thus, a reliable sealing was established around the radiator package to prevent the occurrence of any leakage flows. For each test the radiator module was pre-assembled with screws and attached to the flange of the inlet duct, see Fig. 3 (Table 1).

As stated above, the pressure drop was adjusted using perforated aluminum sheets by varying the hole pattern to simulate different radiator packages. The main design criteria for the pressure drop device was a homogeneous distribution of holes with a

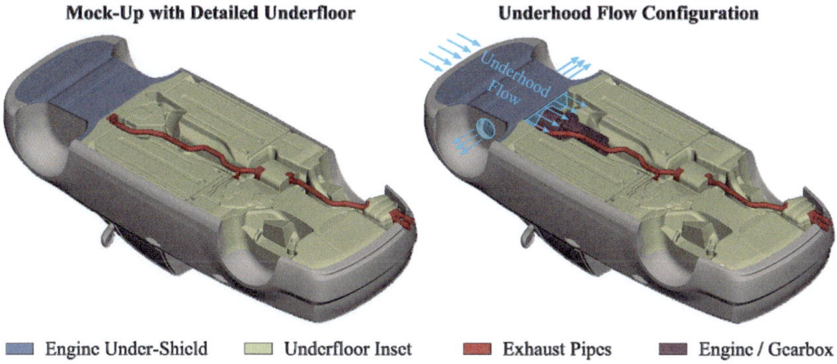

Fig. 1. Comparison of the Mock-Up DrivAer geometry (left) including the detailed underfloor inset with the underhood flow model (right). The geometry is provided by TUM [1].

Table 1. Specifications of the radiator assembly; Δx denotes the thickness of each component.

Part	Δx
Housing Honeycomb 1	15.0 mm
Housing Honeycomb 2	10.0 mm
Honeycomb 1	10.0 mm
Honeycomb 2	10.0 mm
Perforated sheet	0.5 mm
Fan shroud	22.4 mm
Sealing frames (4x)	1.0 mm
$\sum$ Assembly	51.9 mm

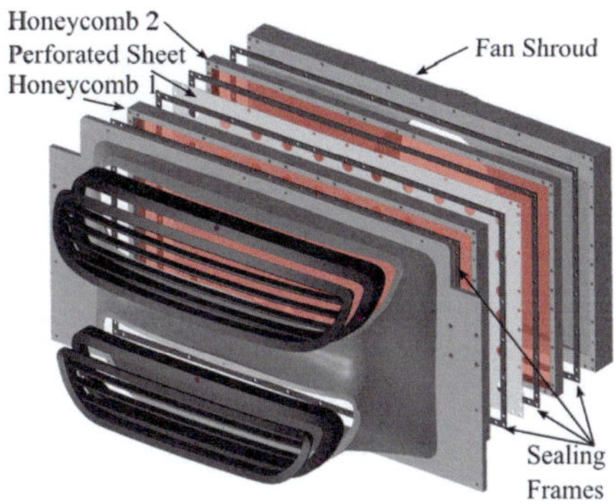

Fig. 2. Comparison of the Mock-Up DrivAer geometry (left) including the detailed underfloor inset with the underhood flow model (right).

Fig. 3. Aluminum fan housing with two flow straighteners and perforated sheets in order to simulate the radiator. Six pressure taps were installed to measure the pressure drop along the module.

Table 2. Specifications of the perforated sheets; A_o/A_c denotes the openness ratio

No. (–)	d (mm)	a (mm)	W_0 (mm)	H_0 (mm)	A_o/A_c (%)
1	4.0	5.0	7.5	7.5	50.3
2	2.5	3.5	8.3	7.8	39.2
3	1.5	2.5	6.3	6.3	28.3
4	1.0	2.0	5.5	6.0	19.7
5	1.0	2.5	7.5	7.5	12.3
6	1.0	4.0	7.5	8.0	4.8

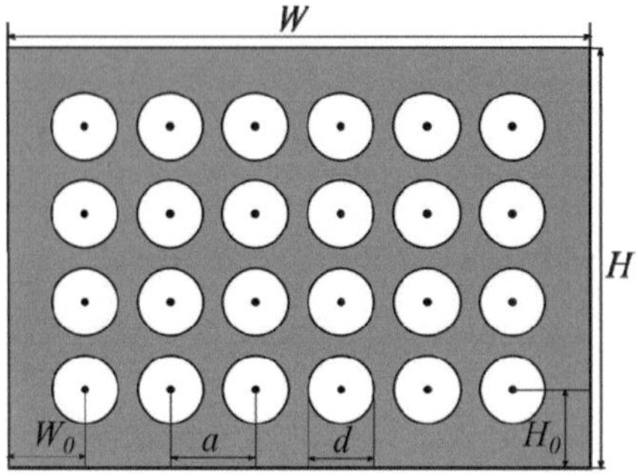

Fig. 4. Sketch of sheet parameters; width W = 180 mm and height H = 335 mm correspond to the fan housing dimensions.

constant and a preferably small diameter on each sheet. The final design was adapted in order to achieve a variation of openness in the range of 4.8% $\leq A_o/A_c \leq$ 50.3%. The specifications for the design of the perforated sheets is presented in Table 2 and Fig. 4 respectively.

The set of perforated sheets was tested in order to evaluate the pressure drop coefficients. The ambient conditions were adjusted to match those from the wind tunnel runs as close as possible:

- Temperature ϑ : 27–28 °C
- Barometric pressure : 980–990 hPa
- Density : 1.13–1.14 kg/m^3

However, additional measurements have been conducted at 22 °C and 32 °C whereas the effect on Δp within the range between 22 °C $\leq \vartheta \leq$ 32 °C was lower than 1%. Figure 5 shows that the selected sheets cover an appropriate range for the pressure drop. Sheet 3 is closest to the curve presented by Wittmeier and Kuthada [10]. However, comparing the coefficients, the curve from Wittmeier and Kuthada [10] returns a higher portion of viscous and hence linear pressure loss (20.72 x) rather than sheet 3 in the present study (9.04 x).

Apart from the variation of the pressure drop along the radiator package the DrivAer model was tested with different rear end shapes (notchback, fastback, estateback). Furthermore, measurements have been conducted for different configurations of the inlets and outlets for the underhood flow. Consequently, the wind tunnel tests comprised measurements with open and closed grille (Fig. 6) as well as open and closed front wheel houses (Fig. 7).

2.2 Experimental Setup

The 40% scale DrivAer model was tested in the WTA with a nozzle cross-section of 4.32 m^2, which provides a reasonable solid blockage ratio of about 10%. Furthermore,

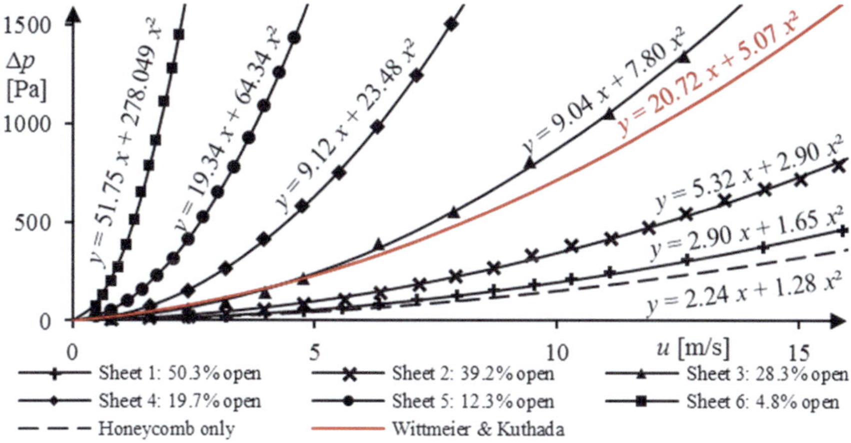

Fig. 5. Pressure drop curves for the 40% scaled DrivAer radiator module including two 10 mm slices of honeycomb and the perforated sheets with different hole patterns. The red curve corresponds to the data presented by Wittmeier and Kuthada [10].

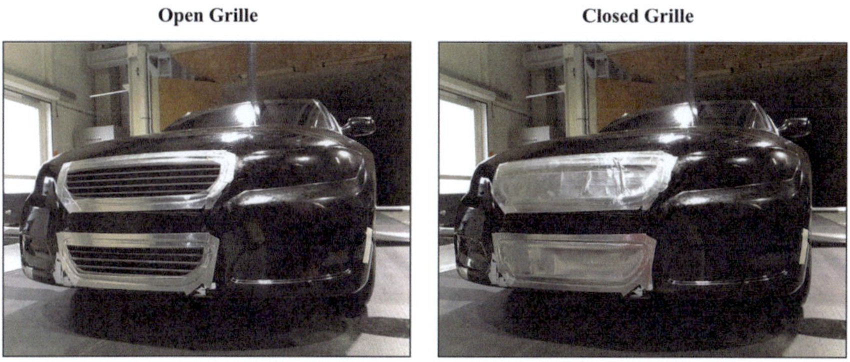

Fig. 6. Open (left) and (closed) upper and lower grille.

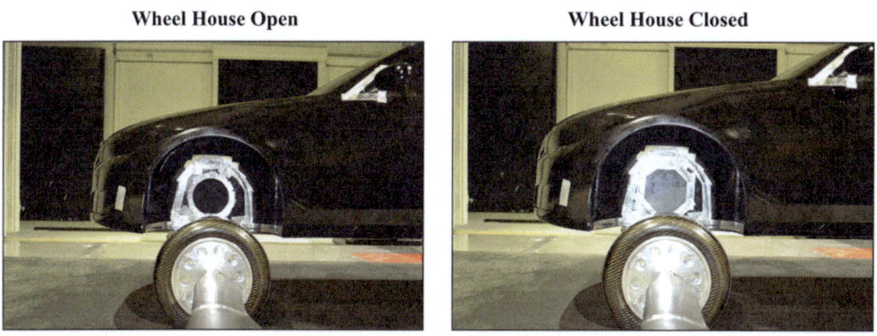

Fig. 7. Open (left) and closed (right) underhood flow outlets inside the front wheel houses.

Fig. 8. Experimental setup of the wind tunnel measurements with moving ground simulation. The 40% scaled model is attached to the top strut, the wheels are attached to four separate wheel struts.

the ground simulation upgrade presented by Mack et al. [3] is expected to be an essential requirement for detailed cooling flow investigations. The moving ground system includes an 1.39 m wide, polyester-based single-belt as well as a passive boundary layer scoop.

Since the moving ground upgrade, a large amount of DrivAer related research work has been under investigation using the 40% scale model in the WTA. Accordingly, experimental data for the Mock-up DrivAer model is provided e.g. by Mack et al. [4], Heft et al. [6], Collin et al. [9] and Haag et al. [14].

For the present study the experimental setup was adopted from Collin et al. [9]. The DrivAer body was mounted to a top strut using an internal six-component balance. The wheels were attached separately to four wheel struts where each wheel was equipped with a one-component force balance to measure the drag of the wheels, see Fig. 8. An alternative model setup without using wheel struts is presented by Miao et al. [8].

Due to the test conditions in summer time the tunnel temperature varied in the range of 22 °C $\leq \vartheta \leq$ 32 °C. The limitations of the maximum moving belt speed applied restrictions to the tests with respect to achievable Reynolds-number as the wind tunnel temperature could not be conditioned to a certain temperature level. Thus, the desired Reynolds-number $Re = 5.2$ million of Collin et al. [9] was not reachable for the entire temperature range. However, we decided to test at a Reynolds-number of $Re = 4.9$ million which corresponded to a testing velocity of 43 m/s $\leq u_\infty \leq$ 45 m/s. An overview of test conditions is presented in Table 3.

Table 3. Wind tunnel test conditions for the 40% scaled DrivAer model with underhood flow in the WTA.

Temperature	ϑ	22…32 °C
Barometric pressure	p_∞	960…970 hPa
Relative humidity	φ	60…80%
Density	ρ	1.08…1.12 kg/m^3
Velocity	u_∞	43…45 m/s
Reynolds number	Re	4.9 million
Dynamic pressure	$q_\infty = q_{NM}$	Nozzle Method (NM)

2.3 Experimental Results

In the following, the results obtained from the wind tunnel tests with the 40% scale model are presented. The presentation of the results contains two general configurations – the Mock-up configuration of the model (after the underhood flow upgrade) firstly, and secondly the tests with open cooling.

2.3.1 DrivAer in Mock-Up Configuration

In the first step, the Mock-up results from the measurements are presented, see Fig. 9. It is worth to mention that the 25% and the 40% scale DrivAer model show differences regarding the model setup. For example, the 40% scale model from TUM was measured with closed, generic wheels whereas the results of the 25% scale model are presented for the 5-spoke wheels. Furthermore, the differences in the experimental setup such as the moving belt systems and, especially, the differences in model support are noteworthy. Consequently, the top and the wheel struts for the 40% scale model are likely to affect the flow around and through the model. On the other side, the openings in the wheel houses are larger and of different shape in the 25% model as the wheels are attached to the model.

However, even though the differences in model setup are significant, the results for both the 25% and the 40% model show a good agreement. The drag coefficient measured on the 40% scaled model was about 0.002 higher than for the 25% scale model. The differences for the front and rear lift coefficient were smaller than 0.008 and 0.012 respectively.

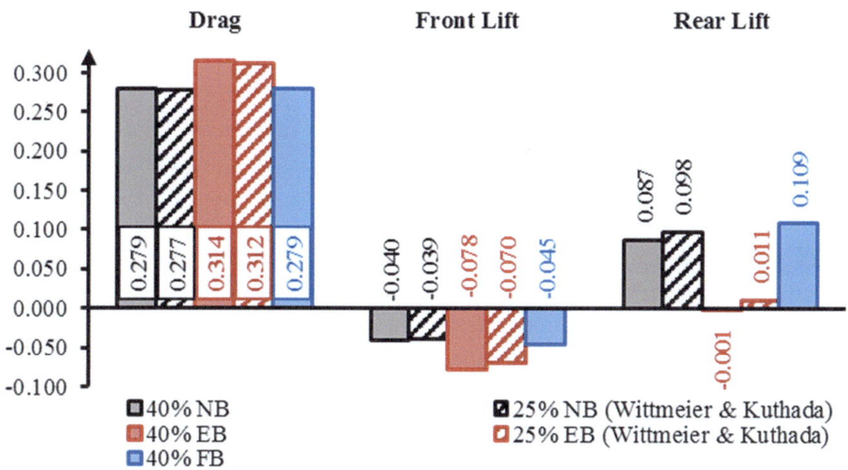

Fig. 9. Measured aerodynamic force coefficients (nozzle method) with moving ground simulation for the 40% scale DrivAer model with underhood flow upgrade in Mock-up configuration compared to the results presented by Wittmeier and Kuthada [10]. The 40% scale model was equipped with closed, generic wheels whereas the results for the 25% scaled model are based on the 5-spoke DrivAer wheels.

2.3.2 DrivAer with Underhood Flow

In the second step, the DrivAer model with three different rear end shapes was tested with cooling flow by varying the pressure drop of the radiator package. Furthermore, the aerodynamic effects of the underhood flow model were investigated with both open and closed openings in the wheel house, see Fig. 7. In all configurations both lower and upper grille were open for underhood flow. Figure 10 provides an overview underhood flow deltas for the underhood flow with open wheel houses and Fig. 11 depicts the results for the model setup with closed wheel houses. Both Figs. 10 and 11 illustrate the effects of the pressure drop in the radiator package on the aerodynamic force coefficients.

Underhood Flow with Open Wheel Houses

In general, the underhood flow effects were higher if moving ground simulation was applied to the test. Especially, the increase of drag is larger for moving ground simulation whereas the effect on front lift was rather small. The biggest cooling drag was measured for the notchback and the fastback as the estateback generally had a smaller cooling drag with ground simulation. This is somewhat inconsistent compared to the results of Wittmeier and Kuthada [10] who measure a larger cooling drag on the estateback ($\Delta C_{D,c} = 0.010$) compared to the notchback ($\Delta C_{D,c} = 0.007$). An explanation for the different results can be again the differences in wheelhouse geometry. Furthermore, the drag increase with underhood flow is expected to be part of the lift induced drag in the wake, which is certainly different for notchback and estateback. Effects from horizontal buoyancy created by a longitudinal static pressure gradient in the WTA were not negligible, too.

Looking at the results for lift, no significant change in total lift was investigated but mainly an effect on the pitching moment was determined for the underhood flow. Consequently, a constant trend of increasing front lift and decreasing rear lift was measured for the open wheel house configuration.

Underhood Flow with Closed Wheel Houses

Looking at the results with closed wheel houses the increase of drag with moving ground simulation for the estateback is rather small ($\Delta C_{D,c,max} = 0.004$) compared to the notchback and the fastback ($\Delta C_{D,c,max} = 0.018$). In addition, the estateback with closed wheel houses was the only configuration which showed bigger cooling drags without moving ground simulation. This different behavior of the estateback was even more conspicuous with respect to rear lift. The configuration of the estateback with closed wheel houses and moving ground simulation showed a remarkable increase of rear lift coefficient of up to $\Delta C_{L,r,c,max} = 0.050$ whereas the rear lift was reduced by $\Delta C_{L,r,c,max} = -0.038$ without moving ground simulation. A connection to the corresponding drag deltas is likely.

Summarizing, the cooling drag is higher if the underhood flow passes the model through both the wheel house openings and the transmission tunnel. In the work presented by Wittmeier and Kuthada [10] this effect was the opposite as they determined the highest cooling drag for the configuration with closed wheel houses. However, a higher cooling drag created by underhood flow through the wheel houses is, according to the authors experiences, reasonable.

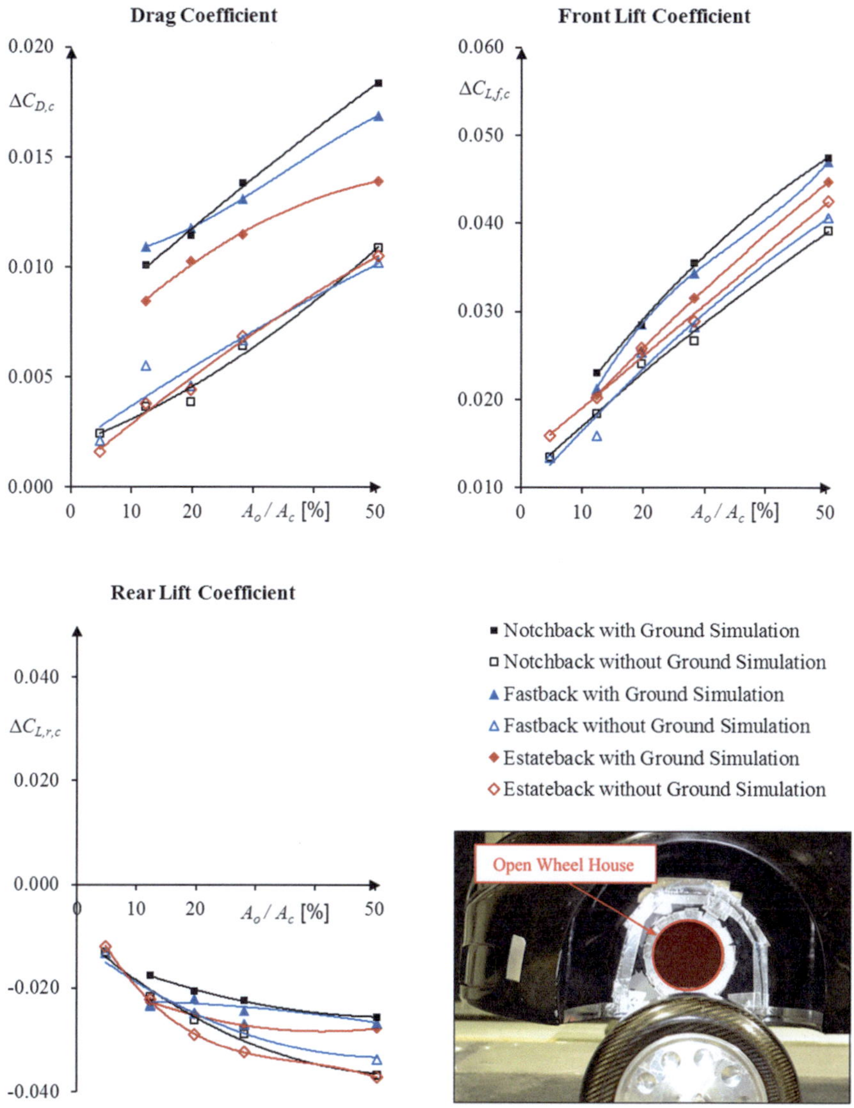

Fig. 10. Effects of underhood flow with open wheel houses on integral force coefficients for drag $\Delta C_{D,c}$ (left top), front lift $\Delta C_{L,f,c}$ (right top) and rear lift $\Delta C_{L,r,c}$ (left bottom). The plots show the results for different perforated sheets with an opening ratio of $4.8\% \leq A_o/A_c \leq 50.3\%$.

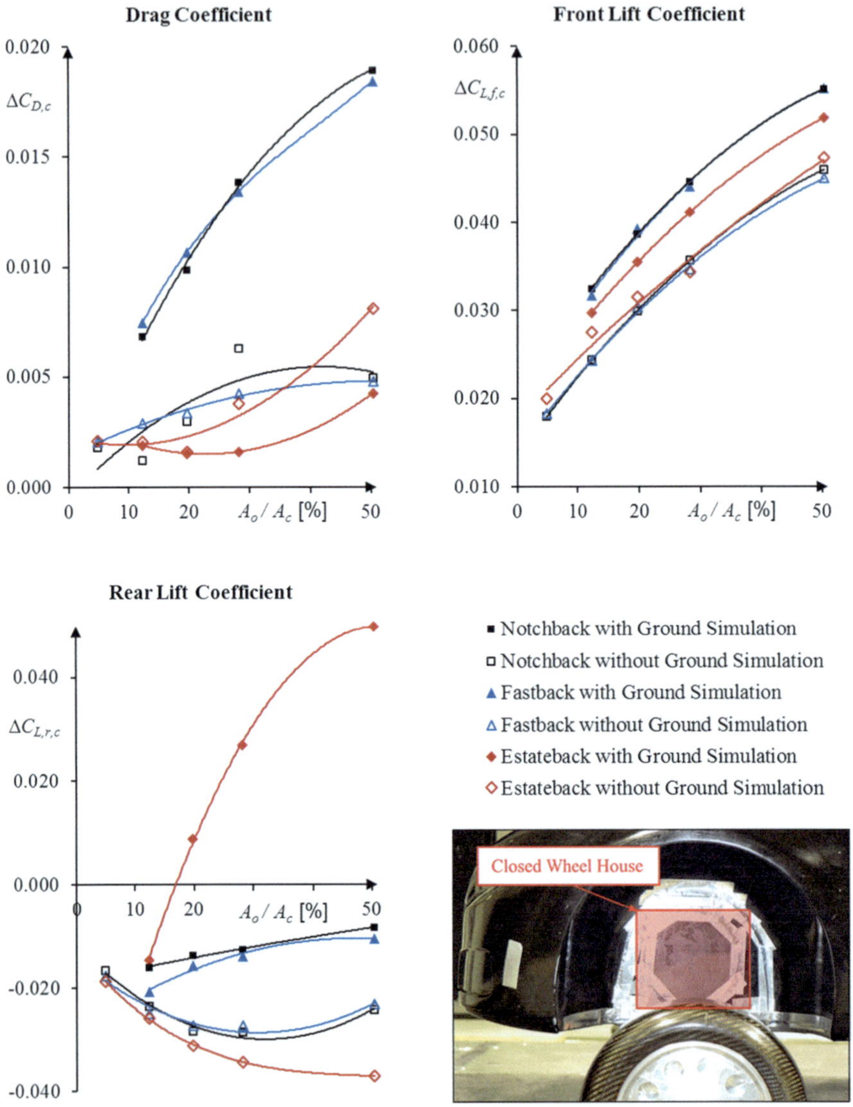

Fig. 11. Effects of underhood flow with closed wheel houses on integral force coefficients for drag $\Delta C_{D,c}$ (left top), front lift $\Delta C_{L,f,c}$ (right top) and rear lift $\Delta C_{L,r,c}$ (left bottom). The plots show the results for different perforated sheets with an opening ratio of 4.8% $\leq A_o/A_c \leq$ 50.3%.

3 Numerical Approach

As a second approach numerical simulations of the DrivAer model were applied in order to compare the effects of underhood flow with the experimental results. In the following, the simulation setup and the results are presented.

3.1 Numerical Setup

In principle, the numerical methods presented by Collin et al. [9] were adopted to run CFD simulations on the open cooling DrivAer model. This includes Delayed Detached-Eddy Simulations (DDES) using the Spalart-Allmaras turbulence model (SA) [15]. The DDES is an unsteady technique based on the Detached-Eddy-Simulation (DES) – which is a hybrid RANS-LES simulation – upgraded with a shielding function to avoid grid induced flow separation ([15–18]). The DDES model is implemented into the open-source CFD toolbox OpenFOAM® version 2.3.x and was optimized in order to develop automotive aerodynamics at the AUDI AG, see e.g. Islam et al. [19]. The near-wall flow was modeled using the Spalding wall function [20]. As presented by Haag [14] the spatial discretization was applied via a second-order central differencing scheme for the convective terms whereas a local, Courant number based blending with a second-order upwind scheme for the velocity was used. The discretization in time was realized using an implicit, second-order backward scheme [9, 14].

The CFD simulations were performed using the 40% scale DrivAer model inside a rectangular shaped domain. The dimensions of the domain were selected in order to achieve a blockage ratio of about 0.084%. Thus, potential open jet interference effects, which might affect the wind tunnel results above, were not considered in CFD.

However, based on the work of Heft [5] we decided to include the model support by top strut as well as the wheel struts as interferences with the DrivAer model are expected, see also CFD results from Collin et al. [9]. Furthermore, the moving belt from the WTA was added to the simulation. An overview is presented in Fig. 12.

In general, the floor of the CFD domain was set to be slip ground. However, to consider the boundary layer profile from the tests [3] as good as possible, the static region between boundary layer scoop and moving belt in the WTA was added to the simulations. Furthermore, the model position on the moving belt is identical to the experimental setup. It is worth mentioning that all geometrical entities were covered by refinement regions to increase the local grid resolution, especially inside the underhood compartment.

The numerical grids were created using the OpenFOAM® integrated, automated meshing tool *snappyHexMesh* and had approximately 125 Million volume cells. After the initialization of the flow field with *potentialFoam* a rather large time step of $\Delta t = 0.0002$ s was selected in order develop the flow field. After 0.5 s the time step was reduced to $\Delta t = 0.00005$ s to increase the accuracy. The physical run time was set to be 2.5 s whereas the flow fields were averaged within the range of 1.0 s to 2.5 s. Thus, an averaging accuracy of the corresponding force coefficients was approximately ± 0.001.

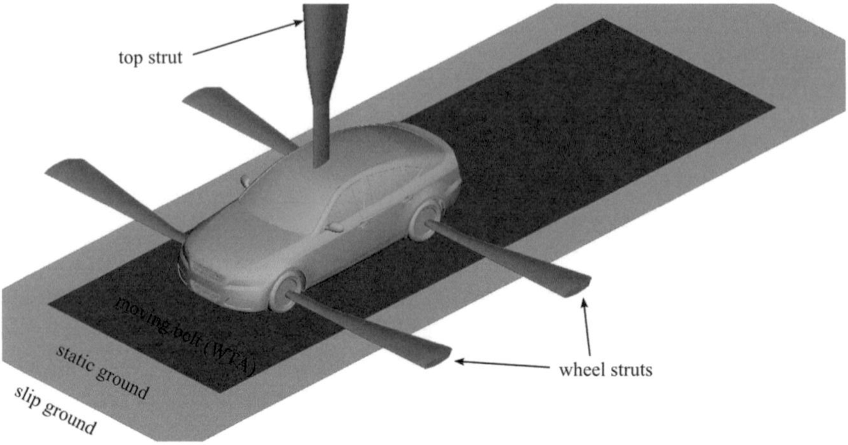

Fig. 12. Geometry setup for the CFD simulations. The domain includes the DrivAer model (here with fastback rear end) with underhood flow as well as the model support by top strut and wheel struts. For the effect of the moving ground simulation the moving belt from the WTA is included, too.

3.2 Numerical Results

For the CFD simulations some DrivAer configurations from Fig. 10 with underhood flow and open wheel houses were selected. This comprised both the notchback and the estateback with different radiator packages and moving ground simulation. The effect of underhood flow on aerodynamic force coefficients, compared to the experimental results, is presented in Fig. 13.

CFD confirms the higher cooling drag $\Delta C_{D,c}$ for the notchback rather than the estateback. The maximum difference between simulation and measurement for $\Delta C_{D,c}$ was ± 0.001 and, hence, within the range of averaging accuracy in CFD. However, the gradient of $\Delta C_{D,c}$ for the estateback turned out to be slightly bigger if the mass flow through was model is increased. The effect on front lift $\Delta C_{L,\mathrm{f},c}$ shows a similar, good agreement with the experimental studies. The influence of underhood flow on rear lift $\Delta C_{L,\mathrm{r},c}$ yielded in bigger differences to the wind tunnel tests (maximum discrepancy ± 0.015). However, CFD still confirms the tendency from experimental studies quite well.

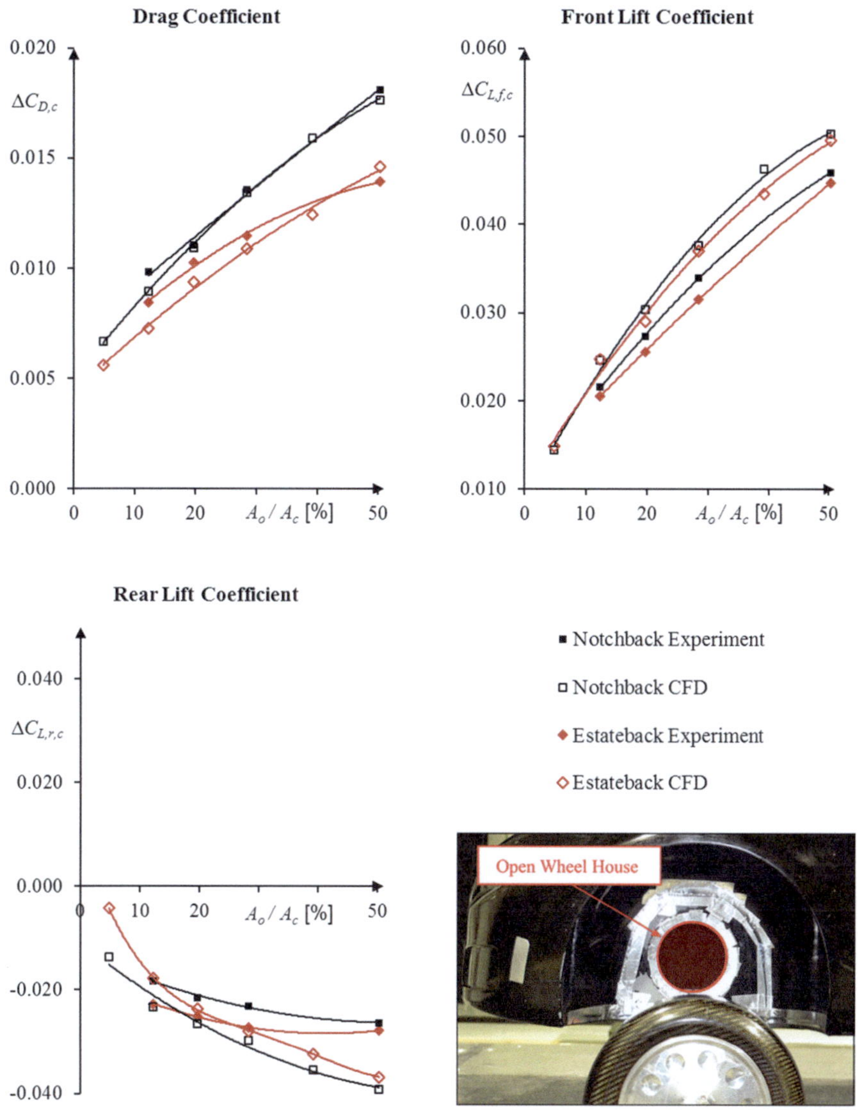

Fig. 13. Comparison of aerodynamic force coefficient from CFD and experimental investigations on underhood flow with open wheel houses. The plots present the integral force coefficients for drag $\Delta C_{D,c}$ (left top), front lift $\Delta C_{L,f,c}$ (right top) and rear lift $\Delta C_{L,r,c}$ (left bottom) for different perforated sheets with an opening ratio of 4.8% $\leq A_o/A_c \leq$ 50.3%.

4 Conclusions

Summarizing, we extracted the influences of underhood flow for the 40% scale DrivAer model of the TUM. This includes results from both experimental and numerical studies for the DrivAer with different underhood flow configurations (cooling flow outlets and radiator packages) as well as different rear end shapes with and without moving ground simulation in the WTA. CFD showed a good agreement with the wind tunnel tests, especially for the cooling drag. However, some general differences to former studies were determined. Thus, in contrary to Wittmeier and Kuthada [10] the cooling drag on the 40% scaled model was higher for the notchback rather than the estateback. Furthermore, Wittmeier and Kuthada [10] recognized a higher cooling drag with closed wheel house openings compared to the closed configuration, which was not confirmed in the present study.

Abbreviations

DDES	Delayed Detached-Eddy Simulation
DES	Detached-Eddy Simulation
LES	Large Eddy Simulation
RANS	Reynolds Averaged Navier-Stokes Equations
SA	Spalart-Allmaras turbulence model
TUM	Technical University of Munich
WTA	Wind Tunnel A (Large Wind Tunnel TUM)

Symbols and Indices

A_c	Closed area of the perforated sheets
A_o	Open/exposed area of the perforated sheets
A_o/A_c	Openness ratio of the perforated sheets
C_D	Drag coefficient
$C_{L,f}$	Front lift coefficient
$C_{L,r}$	Rear lift coefficient
$\Delta C_{D,c}$	Effect of underhood flow on drag coefficient C_D
$\Delta C_{L,f,c}$	Effect of underhood flow on front lift coefficient $C_{L,f}$
$\Delta C_{L,r,c}$	Effect of underhood flow on rear lift coefficient $C_{L,r}$

References

1. Technical University of Munich, DrivAer Model. http://www.drivaer.com. Accessed 9 July 2017
2. Heft, A.I., Indinger, T., Adams, N.A.: Introduction of a New Realistic Generic Car Model for Aerodynamic Investigations, SAE Technical Paper 2012-01-0168 (2012)
3. Mack, S., Indinger, T., Adams, N.A., Unterlechner, P.: The Ground Simulation Upgrade of the Large Wind Tunnel at the Technische Universität München, SAE Technical Paper 2012-01-0299 (2012)

4. Mack, S., Indinger, T., Adams, N.A., Blume, S., Unterlechner, P.: The interior design of a 40% scaled DrivAer body and first experimental results. In: Proceedings of the ASME 2012 Fluids Engineering Summer Meeting (2012)
5. Heft, A.: Aerodynamic Investigation of the Cooling Requirements of Electric Vehicles (Ph.D. thesis). Dr. Hut Verlag, München (2014)
6. Heft, A., Indinger, T., Adams, N.A.: Experimental and numerical investigation of the DrivAer model. In: Proceedings of the ASME 2012 Fluids Engineering Summer Meeting FEDSM2012-72272 (2012)
7. Heft, A.I., Indinger, T., Adams, N.A.: Systematic optimization of the cooling duct system of electric vehicles. In: 31st AIAA Applied Aerodynamics Conference, Fluid Dynamics and Co-located Conferences (AIAA 2013-2520) (2013)
8. Miao, L., Mack, S., Indinger, T.: Experimental and numerical investigation of automotive aerodynamics using DrivAer model. In: ASME. International Design Engineering Technical Conferences and Computers and Information in Engineering Conference, vol. 3 (2015). 17th International Conference on Advanced Vehicle Technologies; 12th International Conference on Design Education
9. Collin, C., Mack, S., Indinger, T., Müller, J.: A numerical and experimental evaluation of open jet wind tunnel. SAE Int. J. Passeng. Cars Mech. Syst. **9**(2), 657–679 (2016)
10. Wittmeier, F., Kuthada, T.: Open grille DrivAer model - first results. SAE Int. J. Passeng. Cars Mech. Syst. **8**(1), 252–260 (2015)
11. Kuthada, T., Wittmeier, F., Bock, B., Schoenleber, C.: The effects of cooling air on the flow field around a vehicle. SAE Int. J. Passeng. Cars Mech. Syst. **9**(2), 723–732 (2016)
12. Matsumoto, D., Haag, L., Indinger, T.: Investigation of the unsteady external and underhood air flow of the DrivAer model by dynamic mode decomposition methods. In: Proceedings of the JSAE Annual Spring Congress, 25–27 May 2016
13. Simmonds, N., Pitmann, J., Tsoutsanis, P., Jenkins, K., Gaylard, A., Jansen, W.: Complete Body Aerodynamic Study of three Vehicles, SAE Technical Paper 2017-01-1529 (2017)
14. Haag, L., Kiewat, M., Indinger, T., Blacha, T.: Numerical and experimental investigations of rotating wheel aerodynamics on the DrivAer model with engine bay flow. In: Proceedings of the ASME 2017 Fluids Engineering Division Summer Meeting FEDSM2017-69305 (2017)
15. Spalart, P.R., Allmaras, S.R.: A One-Equation Turbulence Model for Aerodynamic Flows, AIAA Paper 92-0439 (1992)
16. Spalart, P.R., Jou, W.H., Strelets, M., Allmaras, S.R.: Comments on the feasibility of LES for wings and on a hybrid RANS/LES approach. In: Advances in DNS/LES, pp. 137–147 (1997)
17. Spalart, P.R., Deck, S., Shur, M.L., Squires, K.D., Strelets, M.K., Travin, A.K.: A new version of detached-eddy simulation, resistant to ambiguous grid densities. Theor. Comput. Fluid Dyn. **20**, 181–195 (2006)
18. Spalart, P.R.: Detached-eddy simulation. Ann. Rev. Fluid Mech. **41**, 181–202 (2009)
19. Islam, M., Decker, F., de Villiers, E., Jackson, A., Gines, J., Grahs, T., Gitt-Gehrke, A., Comas i Font, J.: Application of Detached-Eddy Simulation for Automotive Aerodynamics Development, SAE Technical Paper 2009-01-0333 (2009)
20. Shih, T.H., Povinelli, L.A., Liu, N.S., Che, K.H.: A generalized wall function for complex turbulent flows. In: Proceedings of the 8th Aerospace Science Meeting and Exhibit sponsored by the AIAA (2000)

Potential of Porsche Reference Cars for Aerodynamic Development

Francesca Cogotti(✉), Michael Pfadenhauer and Thomas Wiegand

Porsche AG, Porschestrasse 911, 71287 Weissach, Germany
francesca.cogotti@porsche.de

Abstract. The new Porsche wind tunnel was built in three years from 2011 and 2014. In the meantime, Porsche decided to build two reference cars, so as to be able to compare the pressure distribution on the vehicle surface, both on the road and in the wind tunnel.

The concept of having a reference car to compare results on the road and in the wind tunnel is not new. In the case of Porsche, the choice of the vehicles was guided by two requirements, i.e. to have a car shape typical of the Porsche family and to produce different blockage effects.

The cars were equipped with the best available instrumentation, which was carefully calibrated in order to insure accuracy and repeatability.

Multiple road tests and measurements in various state of the art wind tunnels, with different moving ground simulation systems and different geometries, were then carried out.

As a result of the post processing of these data, it has been possible to recognize interference effects in the various wind tunnels.

Porsche has then used these vehicles in the calibration phase of the new wind tunnel, so as to reach a flow quality, in terms of time-averaged values, that is as close as possible to conditions experienced on the road.

Furthermore, thanks to the interchangeable moving ground system in the new Porsche wind tunnel, the reference cars were able to show clearly, for the same wind tunnel interferences, the differences in the pressure distribution caused by the two different moving ground systems.

In addition, thanks to the high-quality 5 belt system, it was possible to measure the c_{D_vent} (ventilation drag), necessary to improve the prediction of Fx in coast-down tests. A comparison between wind tunnel and coast-down results was carried out and will be shown.

1 Introduction

In modern road car development projects aerodynamics is becoming increasingly more important. More demanding targets for overall vehicle performance and for fuel consumption and emissions require an ever-increasing expenditure of effort. Furthermore the new WLTP rules promote the need for the development of test conditions to be as close as possible to the real conditions on the road. For instance, wind tunnels without some type of moving ground simulation are nowadays unacceptable.

J. Wiedemann (ed.), *Progress in Vehicle Aerodynamics and Thermal Management*,
https://doi.org/10.1007/978-3-319-67822-1_15

Hence Porsche decided to invest not only in a new aeroacoustic wind tunnel facility, but also in two reference cars, to be used to understand the differences and fill the gap between the results of wind tunnel tests, CFD simulations and road tests.

The new Porsche wind tunnel project was focused on some specific requirements, important for car development at Porsche.

The first requirement was the capability to reach a wind speed of 300 kph in the standard wind tunnel configuration. An interchangeable nozzle system, with two different exit sections, for example of 25 m^2 and 18 m^2 as shown in [1], was taken into consideration and then rejected as impractical. The final decision, seen as the best trade-off, was to build a nozzle with a cross section of 22 m^2.

The second requirement was to be able to have on demand the possibility to choose between two different moving ground systems (a 5 or a 1 belt system, both up 300 kph), depending on the vehicle project under development or on the customer needs.

Both these requirements are important in a company like Porsche, which develops cars characterized by extreme performance with top speeds of up to 340 kph, and are heavily involved in sporting competitions such as the WEC (World Endurance Championship) with GT and LMP cars.

These and the other project requirements were fully achieved and verified in the commissioning phase of the empty wind tunnel. The details of the new Porsche wind tunnel were presented at the FKFS Symposium in 2015 [2].

It was decided that it was also very important to verify that the wind tunnel performance remains excellent not only when the test section is empty but also when vehicles of different shape and size are in the flow.

To do that, in the year 2010 Porsche started a parallel project to build a first reference car, equipped with all the necessary on-board instrumentation so as to enable comparison of surface pressure distributions measured on the road or in the wind tunnel, and be able to independently measure all the relevant test conditions during such tests.

The first reference car was a sports car, and was ready in 2012. The vehicle was then tested in several state-of-the-art wind tunnels, that is to say wind tunnels built according to the most recent achievements in the design and in the measuring systems, and on different road test-tracks over a period of about 3 years, in order to carefully and thoroughly calibrate it. In this first part of the project, road-tests were focused on measurement at constant velocity, mainly at 140 kph.

Following the experience with the first car, Porsche decided to prepare a second reference car, an SUV, in order to have a second vehicle able to produce a different blockage effect in the wind tunnel.

Both these vehicles were used to check the calibration of the new Porsche wind tunnel. Using these two reference cars, Porsche was able to perform a final refinement, so as to improve the flow quality, the repeatability of the results and specifically to tune the wind tunnel to deliver flow conditions as close as possible to those experienced during road tests.

2 Overview of the Reference Cars

The concept of having a reference car to compare results on the road and in the wind tunnel is not new, see [3, 4].

However, Porsche set up two vehicles, a sports car and a SUV, so as to have different vehicles with a typical Porsche shape but two different blockage effects.

The main dimensions of these vehicles are listed in Table 1.

In order to ensure the repeatability of the test results on the road, and in the different wind tunnels, some modifications were carried out on the two vehicles.

For instance:

- Flexible wheel spoilers and air-dams were replaced by hard add-on parts.
- For the sports car, the original surface of the rear spoiler characterized by air slots was replaced by a flat surface, so as to avoid critical flow separations.
- At the rear of the underbody, a plate was added, so as to have a surface on which to measure the centerline pressure distribution.

Table 1. Reference car dimensions

Car details	Sports car	SUV
Vehicle frontal area (m^2)	2.0	2.8
Vehicle length (m)	≈4.43	≈4.85
Vehicle height (m)	≈1.3	≈1.7

The cars were equipped with dedicated state-of-the-art, modern vehicle-mounted, instrumentation to perform the following measurements (see Table 2 and Fig. 1):

- Pressure distribution all around the vehicle, with pressure ports on the car upper and lower sides, on the vehicle underbody and within the wheel housings.
- Vehicle ground speed acquired with a "Correvit" laser measuring system.
- Vehicle ride heights, by 4 laser-based measuring systems, located in the underbody area.
- Tire temperatures acquired with 4 pyrometers, to check the temperatures in the road-tests as well as in the wind tunnel during tests performed with different belt systems.
- Rpm of 3 fans, 2 behind the cooler and one in the engine bay.

Table 2. Instrumentation on board

Instrumentation on board	Sports car	SUV
Pressure ports on the vehicle surface (state-of-the-art dedicated pressure scanners)	Approximately 340	Approximately 180
Wind velocity (Pitot) and belt/road velocity (Correvit)	Measured	Measured
Ground clearance (laser)	Measured (×4)	Measured (×4)
Tire temperature, road temperature (pyrometer)	Measured (×5)	–
Fan rpm	Measured	Measured
Oil temperature	Measured	–
Fan to simulate exhaust	Included	–

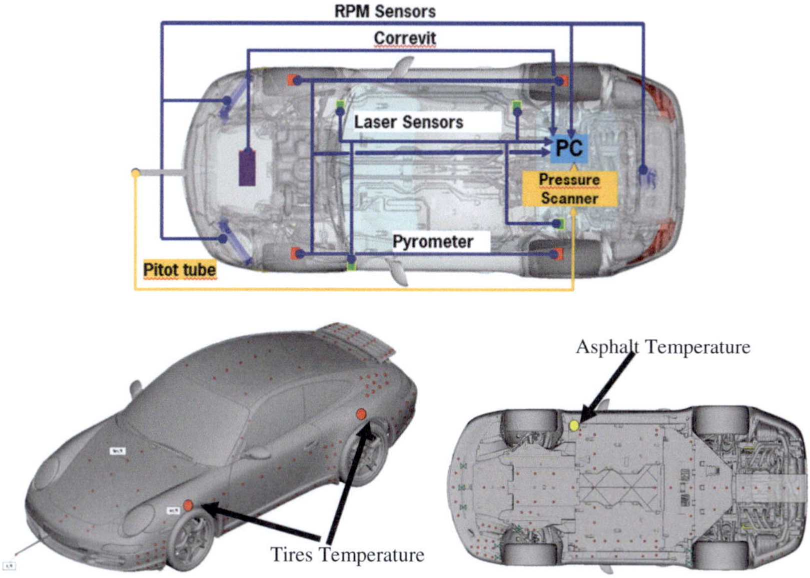

Fig. 1. Schematic showing the onboard instrumentation and the locations of the pressure ports.

2.1 Pressure Sensors on the Vehicle and Vehicle Calibration

The first reference vehicle, the sports car, was prepared with about 340 pressure ports. The ports are distributed on the car top surface, on the underbody, in the wheel arches and on the radiator, on the front and on the back. On the second vehicle, fewer pressure ports (180) were used, distributed on one side, in the most important areas such as the top surface, underbody and wheel arch.

Pressure scanners were chosen to be as accurate as possible, in particular at 140 kph. The measuring range is ±2.5 kPa, with an accuracy of ±2.5 Pa.

As it is known

$$c_{p(x)} = \frac{p_x - p_\infty}{q_\infty}$$

Where:

p_x is the static pressure at a pressure tapping
$p\infty$ is the free-stream static pressure
$q\infty$ is the free-stream dynamic pressure

The main difficulty on the road tests is to establish where to measure $p\infty$.

However if one finds a point on the vehicle surface where $c_p \approx 0$ and that doesn't change with wind speed, not in presence of small yaw angles or small changes of standing heights, one can take the static pressure at this point as the free-stream static pressure ($p_{ref} \approx p_\infty$)

Then one has:

$$c_{p(x)}* = \frac{p_x^*}{q^*} = \frac{p_x - p_{ref}}{p_{tot} - p_{ref}}$$

Where

$p_x^* = (p_x - p_{ref})$ is the static pressure at a generic pressure tapping measured on the reference car

$q^* = (p_{tot} - p_{ref})$ is the dynamic pressure at the Pitot head

As explained in [3], four points on the vehicle need a calibration and have to be carefully selected in order to measure the flow speed and the flow angle while driving on the road or testing in the wind tunnel (Fig. 2):

P1. p_{ref}, the reference pressure on the bonnet, where the c_P should be circa zero

P2. p_{tot}, the total pressure as measured by the probe on the front of the vehicle

P3-P4. on the right and on the left side of the vehicle, to determine the yaw angle

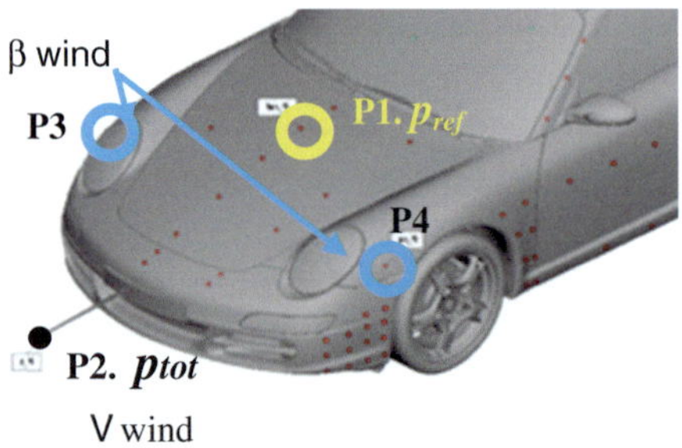

Fig. 2. The most important pressure ports on the car. Once calibrated they are used to determine the wind velocity and yaw angle

The validity of our choice was checked by changing several parameters, such as flow velocity, small yaw angle, ride height, vehicle configuration, and moving ground simulation. Figure 3 shows the stability of the pressure at the reference point. All the tests show a change of the reference pressure in a range of ±3 Pa at 140 kph, which means a delta c_P of +/0.003.

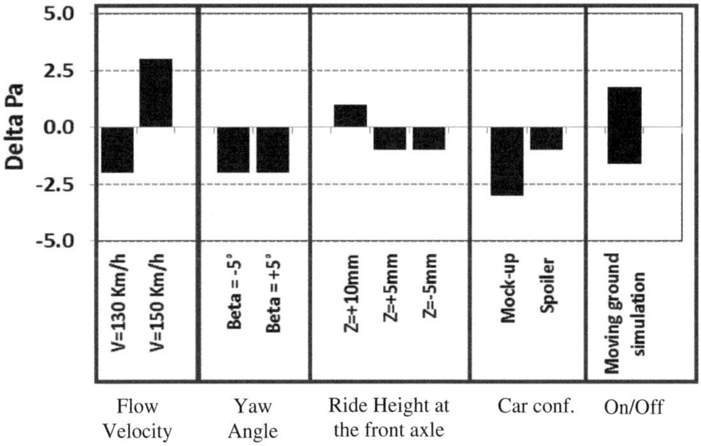

Fig. 3. Sensitivity of p_{ref} in case of changes to the base configuration at 140 kph

2.2 Vehicle Velocity

Velocity is measured by a GPS system on the vehicle and by a "Correvit" laser measuring system located in the trunk lid. The Laser has been protected with a glass in order to avoid damages during the road tests. The influence of the glass in the measurements have been proofed and it is negligible (about 0.05 kph). The Correvit has been used also in the various wind tunnels to check the belt velocity.

2.3 Vehicle Ride Height

Ground clearance is measured by four laser sensors located on the vehicle underbody.

The ride height is stored and checked with errors of less than ±0.1 mm.

2.4 Temperature

To protect the static pressure tubes during road tests, several pyrometer temperature sensors are positioned in the engine compartment enabling critical areas to be monitored, and harmful temperatures avoided during the road-tests.

Additionally, to monitor tire temperatures on the road tests or during wind tunnel tests with different belt systems, four pyrometers were located in the wheel arches.

For the coast-down tests, an additional dedicated sensor has been added to measure the road surface temperature in order to compare this with the data available from the weather station.

3 Overview of the Main Results Measured on the Road Tests and in Various State-of-the-Art Wind Tunnels

The calibration-phase of the reference cars in various test conditions required a large amount of time, but it was the most important and productive phase of this project. In fact the post-processing of all the measurements with the two vehicles, made in different wind tunnels with different blockage effects, different moving ground systems or on different proving grounds, produced an important secondary result in identifying the interference effects in the various wind tunnels.

Table 3. Overview on the several tests carried out

Tests		Sports car	SUV
WIND TUNNEL* * The new Porsche WT is not included in this table	Without moving ground simulation	1	1
	With moving ground simulation (5 belt)	4	2
	With moving ground simulation (1 belt)	2	1
CLIMATIC WIND TUNNEL	With rollers	1	–
ROAD TESTS Constant speed	Test track	PAG Lahr Ehra	Ehra
COAST-DOWN	Test track	IDIADA	–

The state-of-the-art wind tunnels were chosen first of all in accordance with the targets of the new Porsche wind tunnel, that is to say they offered the possibility of testing with different belt systems, and if possible with a nozzle equal or bigger than 22 m^2.

In addition, tests were also carried out in facilities such as the old Porsche wind tunnel and a climatic wind tunnel, so as to have a link with the facilities where Porsche develops or has in the past developed vehicles (Table 3).

3.1 Main Wind Tunnel Effects

Data measured in the wind tunnels were compared with the data measured on the road. The first and simplest comparison is of the pressure distribution on the vehicle centerline. The closer the deltas are to zero, the better the performance of the wind tunnel (See Fig. 4).

The main effect of the wind tunnel on the vehicle surface pressure distribution is best shown using the example of the data measured in a climatic wind tunnel (CWT, red line), where these effects are clearly visible due to the limited nozzle dimensions (about 5 m^2).

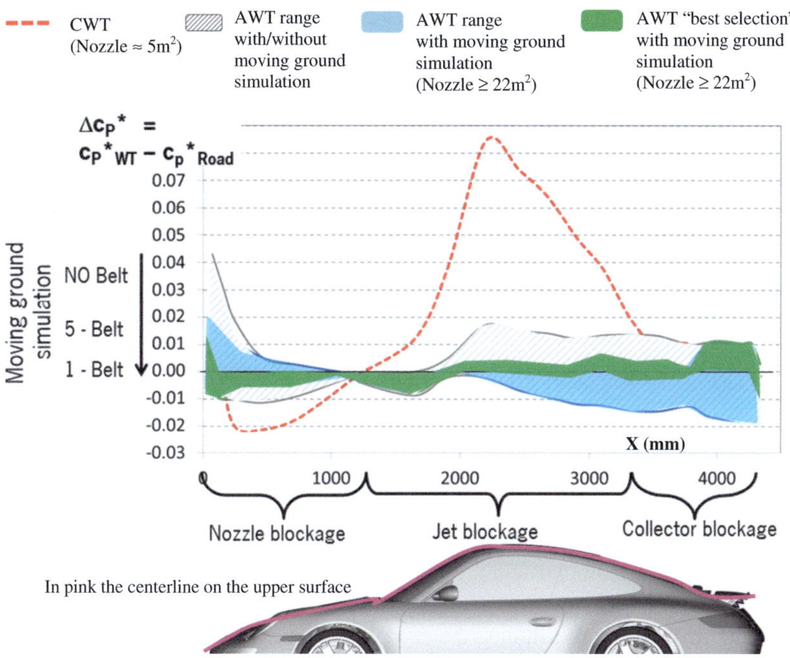

Fig. 4. The delta between the c_{P*} measured in the wind tunnel and on the road on the centerline (Y = 0 in pink).

Wind tunnel effects are clearly reduced in "aerodynamic" wind tunnels (AWT, grey area). In order to select a flow condition closer to that of the new Porsche wind tunnel, the range picked out in blue shows only the measurements carried out in wind tunnels with a nozzle section equal or bigger than 22 m^2 and with moving ground simulation (AWT, blue area).

The influence of the moving ground simulation can clearly be seen in the first pressure points on the bonnet. The pressure delta to the road in a wind tunnel without any moving ground simulation is obviously bigger than in a wind tunnel with moving ground simulation. The two wind tunnels with a single belt show a further improvement in comparison with the 5 belt system.

One wind tunnel showed an unexpected flow condition and it was decided to remove this wind tunnel from the "best selection" (shown as the green area in Fig. 4). More details are available in Sect. 3.2.

For the sports car, the difference between the "best wind tunnels" and the road shows a c_P range of ±0.01. This green area in Fig. 4 was the target for the new Porsche wind tunnel. The "best wind tunnels" have a pressure gradient, that is flat in the area of interest (±3 m), see Fig. 5, therefore we can assume that the interference effects, if any, are very small and are not affecting the measuring method.

The same was done for the SUV and a similar target "green area" defined.

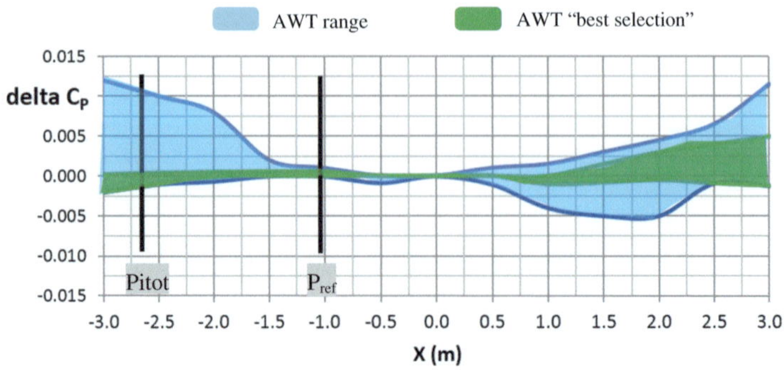

Fig. 5. Range of the longitudinal static pressure distribution in the "AWT" and "AWT best selection".

3.2 Wind Tunnel Velocity and Aerodynamic Coefficients

From the comparison between different wind tunnels, it was interesting to notice the difference between the official wind tunnel velocity and the flow velocity measured by the vehicle.

The state-of-the-art wind tunnels have a pressure gradient in the empty test section that is close to zero in the area of interest for the vehicle.

To compare the velocity shown from the Pitot tube on the vehicle, it is important that between the Pitot and the reference static pressure on the bonnet, there are no wind tunnel effects. This can be checked by doing reference to the pressure gradient in the empty test section.

Figure 5 shows the "correctness" of the chosen wind tunnels. The longitudinal static pressure delta is between −0,005 and 0.012 in the vehicle area (±3 m). As in Fig. 4, the blue area includes the pressure distribution in the wind tunnels with a nozzle section greater than 22 m^2 and with moving ground simulation. The "best selection" are shown in green.

The vehicle reference static pressure (p_{ref}) measured vs. the wind tunnel reference is a bit different in each wind tunnel. However, if the pressure gradient in the empty test section is approximately constant in between the X locations of the vehicle reference static pressure and the Pitot, this difference is negligible and the wind speed shown from the Pitot is not affected by this difference. The X position of the Pitot and of the p_{ref} are shown in Fig. 5

The analysis of the data has shown that for two wind tunnels the difference in the velocities was important. It was interesting to see that these effects had different causes in each case. At 140 kph, between the fastest and the slowest wind tunnel there was a velocity difference of about 9 kph.

In wind tunnel #5, the c_D of the car was about 30 points higher than in the other wind tunnels (see Fig. 6), however the vehicle pressure distribution compared well with the road tests (see Fig. 7) and the longitudinal pressure distribution in the empty wind tunnel has a maximum delta c_P of ±0.001 in an area from X ∓3 m.

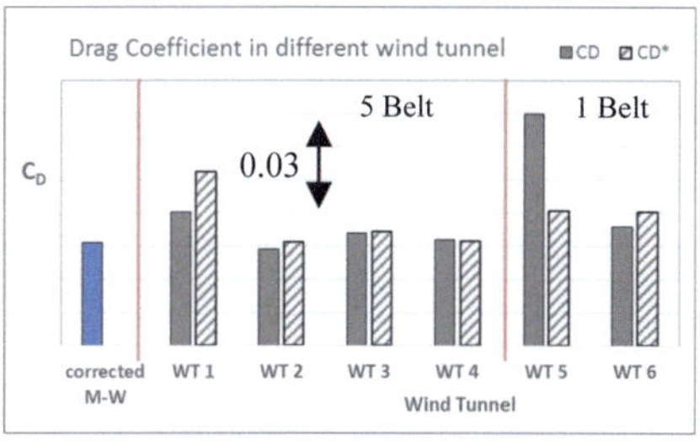

Fig. 6. The drag coefficient in different wind tunnels. c_D* is the value of the c_D recalculated with the real velocity. In blue the value by using the Mercker-Wiedemann correction formula [5–9].

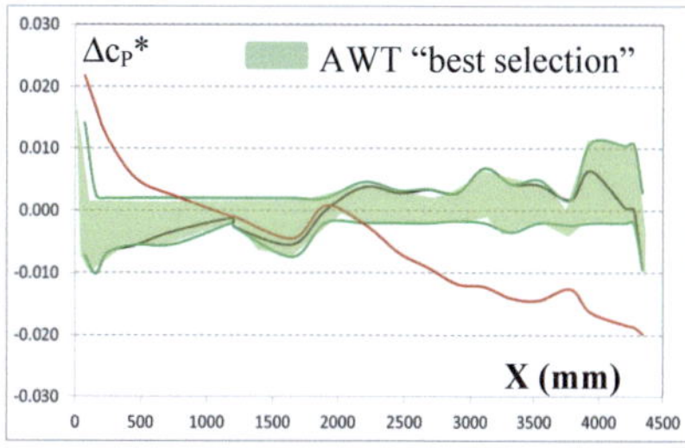

Fig. 7. The delta c_P* in comparison with road tests on the centerline. WT#5 (black line) is in the "green area", the area of the "best wind tunnels". Wind tunnel #1 (red line) is not. Delta c_P* = 0 means the wind tunnel c_P* is the same as on the road.

The Pitot was helpful to show that the cause of the higher c_D value was a higher velocity in the wind tunnel in comparison with the official wind tunnel calibration. When the c_D is recalculated with the velocity measured by the Pitot, the differences between this WT and the others are much lower.

After this test session, the wind tunnel calibration was checked by the operators and it was confirmed that for several reasons the wind tunnel velocity needed a new calibration.

Instead, in wind tunnel #1, the c_D value was close to those of the other wind tunnels, however once again the Pitot showed a difference in the flow velocity. This time the real wind velocity was lower than the official one. The difference in static pressure in the empty wind tunnel between the Pitot and the p_{ref} is not of sufficient magnitude to explain it. The surface pressure distribution did not correlate well with the other wind tunnels (red line) and this was not improved with increased wind speed. The difference in the base pressure should have generated a higher c_D and this did not happen due to the lower real velocity. For this reason we decided to remove this wind tunnel from our comparison.

4 Final Tuning of the New Porsche Wind Tunnel

As reported in Chap. 1, the aim of the new Porsche wind tunnel was not only to satisfy the specification requirements in the empty test section, but also to have a pressure distribution on the surface of the test vehicle as close as possible to that measured on the road and at least comparable to other of today's state-of-the-art wind tunnels.

After the calibration of the empty test section, the first measurements were conducted using the two reference cars. Results are shown in Fig. 8a and 8b. The pressures along the centerline on the upper part of the vehicle and on the rear end show that pressure values were not as close to the road as in the other wind tunnels (black line in Fig. 8a).

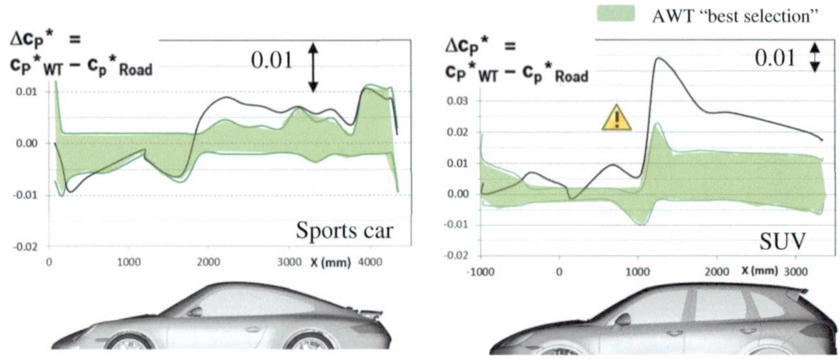

Fig. 8a. The Delta c_P* on the car centerline (Y = 0) in the initial wind tunnel configuration (black line = Porsche new wind tunnel). On both the cars, the difference to the road is bigger than in other state-of-the-art wind tunnels (green area).

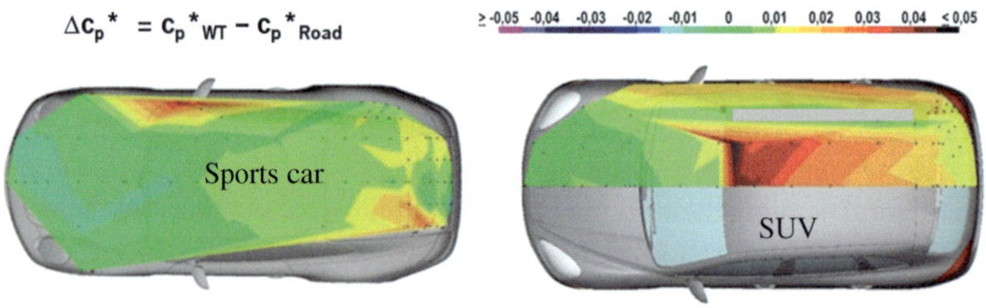

Fig. 8b. The Delta c_P* on the car surface in the initial wind tunnel configuration (green means as on the road)

Although the pressure gradient in the empty test section was flat, further optimization of the pressure distribution on the vehicle was pursued.

Different wind tunnel modifications have been tested to achieve the specifications. Changes were mainly focused on the nozzle and on the collector.

The best result was achieved with a nozzle extension (pink line in Fig. 10).

With a nozzle extension of 1.3 m, the target pressure distributions on the reference vehicles and the specifications for the pressure gradient in the empty test section were both achieved. As shown in [2], the specification for the empty test section was a maximum delta c_P of ±0.002 in an area from X −3.5 to +4.5 m.

Figure 9 shows the pressure distribution in the empty test section, without and then with the nozzle extension. A car silhouette is superimposed on the diagrams to show the region where it is of greatest importance that the pressure gradient remains as flat as possible.

The final wind tunnel configuration and details of its flow quality has been presented at the VDI Wissensforum in 2016 [10].

Figure 10 shows the changes in the pressure distribution on the centerline and Fig. 11 the comparison between the final wind tunnel configuration and the road. With the nozzle extension the Porsche wind tunnel is also in the range of the "best wind

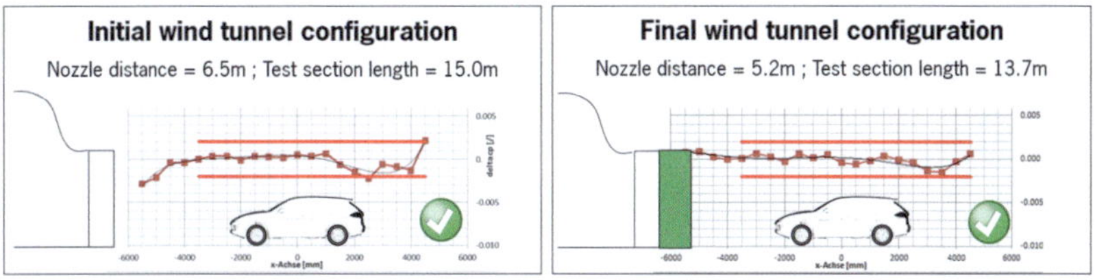

Fig. 9. Longitudinal static pressure distribution in the empty test section in the new Porsche wind tunnel at Z = 0.75 m.

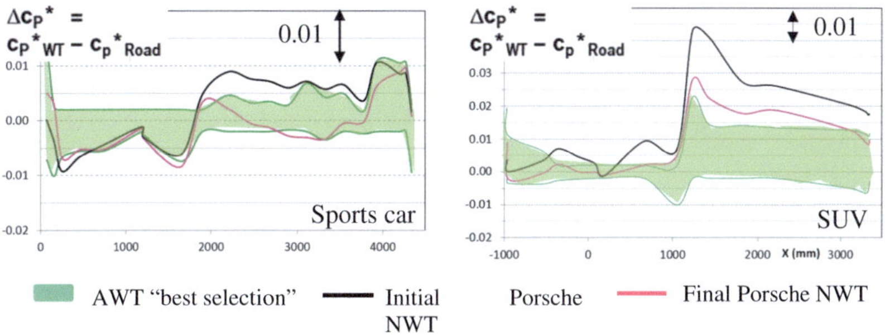

Fig. 10. The delta c_P* on the centerline of the reference vehicles in the initial (black line) and final (pink line) wind tunnel configuration.

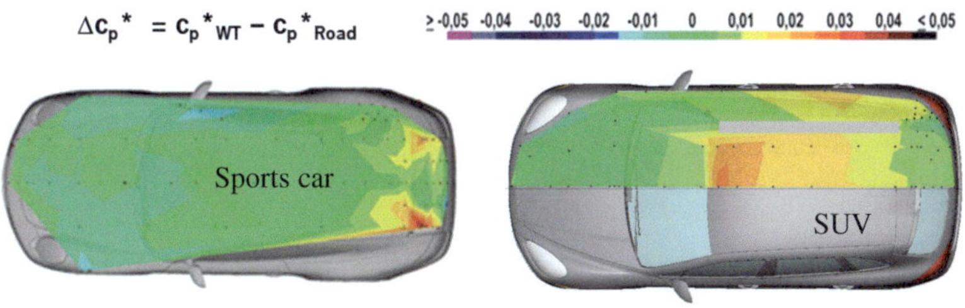

Fig. 11. The delta c_P* on the surface of the reference vehicles in the final wind tunnel configuration.

tunnels". In particular on the SUV, the pressure coefficient on the base was decreased by about 0.010 (see Figs. 12 and 13). The drag coefficient of this car was thereby increased by around 5 counts.

The areas of the car where the surface pressures compared poorly with the road tests were those where the static pressure in the wind tunnel was higher than on the road. In areas such as the car roof, where the surface pressures are more positive than on the road, it means that the flow velocity in the wind tunnel was slower than on the road.

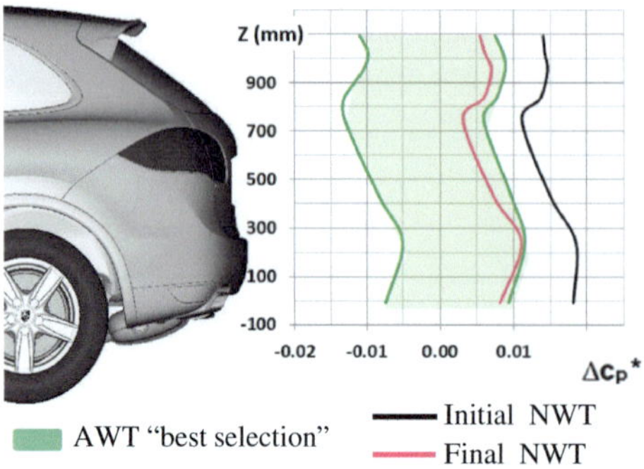

Fig. 12. The delta c_P* on the centerline on the back of the SUV in the initial (black line) and final (pink line) WT configuration.

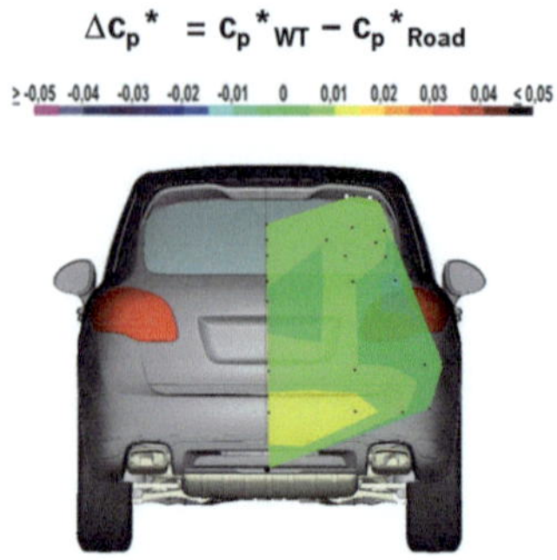

Fig. 13. The delta c_P* on the surface of the SUV in the final wind tunnel configuration

The relationship between the nozzle cross-section and the vehicle-to-nozzle distance was not optimal and the jet expansion (Fig. 14) around the vehicle was the main reason for these differences in the pressure distribution. Furthermore, this excessive expansion also produced an instability in the flow that caused poor repeatability of the results. In the literature [11] information is available about the optimum length of the test section in relation to the nozzle section.

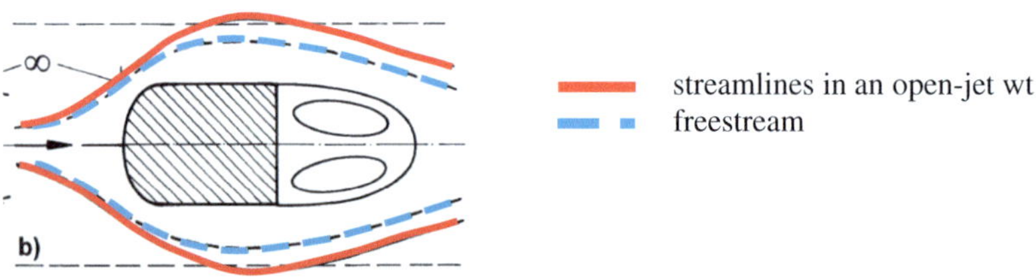

Fig. 14. Schematic of jet expansion [11]

In 2015, whilst this project was underway, very little published information was available concerning the relationship between the vehicle-to-nozzle distance and the nozzle cross-section.

In the final wind tunnel configuration the vehicle distance from the nozzle was reduced from 6.5 m to 5.2 m. A similar study, published in 2016, see [12], which sought to optimise a wind tunnel layout in CFD prior to construction, shows similar results.

As anticipated, a second advantage given by the nozzle extension is that the repeatability of the measured aerodynamic coefficients has been improved.

Figure 15 shows the drag coefficient of the sports car in several tests conducted over a period of one year, from the first wooden “concept” nozzle extension to the final version constructed in steel. Repeatability is now extremely good and the discrepancy in the measured values is in a range smaller than 0.001

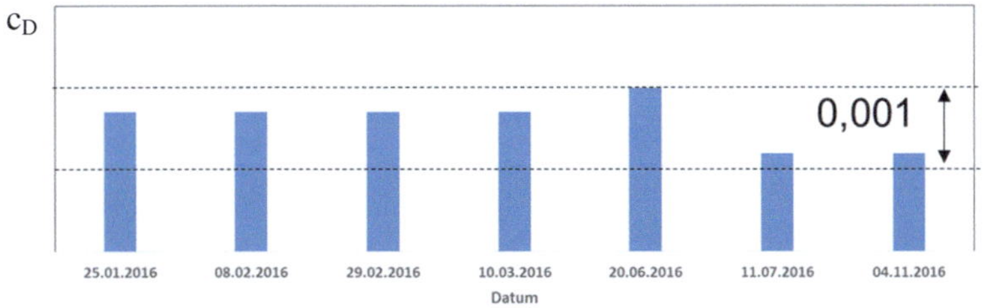

Fig. 15. The c_D of the sports car, measured in several wind tunnel sessions over a period of one year and showing the high level of repeatability achieved

5 Comparison, 5 Belts to 1 Belt System

The new Porsche wind tunnel is the only wind tunnel in the world where it is possible to modify the moving ground simulation system from a 5 belt system to a single belt system in less than 3 h. Information about the used configuration in the two systems are reported in Table 4.

Table 4. Belt systems configurations compared in this work.

	5 belt system (standard)	1 belt system
Center belt	9 × 1.1 m	9 × 3 m
Wheel rotation unit	Front 0.24 × 0.28 m Rear 0.24 × 0.34 m	–

When measured using the single belt system, the vehicle surface pressure distribution is closer to that measured on the road. This effect is markedly more apparent for the sports car than the SUV, where the difference is negligible.

In particular the differences in the pressure distribution are seen on the rear end of the car and on the underbody.

Figure 16 shows the difference in the pressure distribution on a Z-line on the back of the vehicle in comparison with the road. The "green area" shows the results of the "best wind tunnels" with a 5 belt system. In the "violet area" are the wind tunnels with a single belt. The Porsche wind tunnel is in both cases comparable with the other "best wind tunnels". The pressure on the rear end of the car when measured on the single belt is closer to the road in comparison with the 5 belt systems. This improvement is of the order of 0.015 in c_p.

One of the disadvantages of the single belt is that there is no means of exactly controlling or measuring the ride height. For the reference cars the ride heights have been measured using the lasers, to ensure that pressure distributions are only compared between identical ride height configurations.

The advantage of the single belt in terms of better flow quality around the car is known and visible in Fig. 17. For super sports cars or racing cars with reduced ground clearance a single belt system has to be recommended. Its main value is to compare

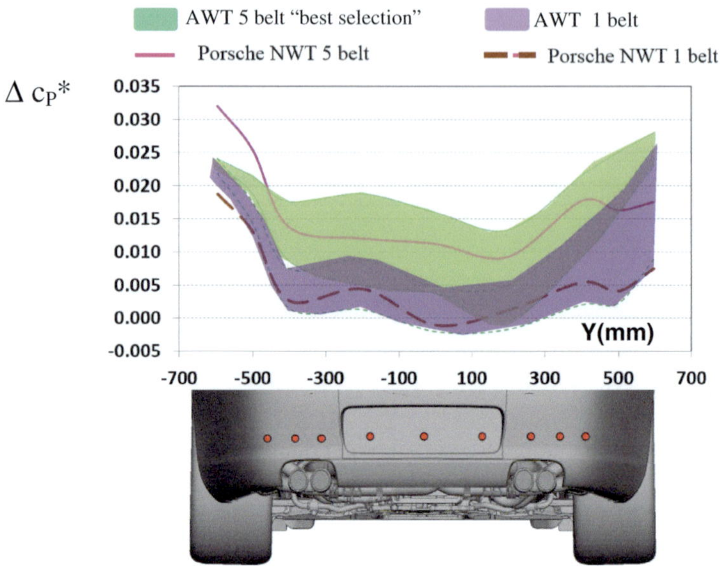

Fig. 16. Delta c_P* between wind tunnel and road on a Z line on the rear end of the sports car.

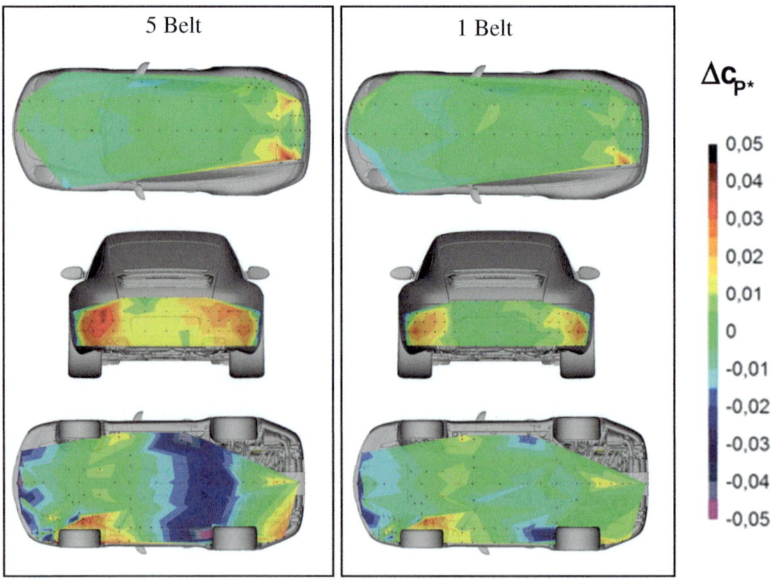

Fig. 17. Delta c_P* on the surface of the sports car reference vehicle measured on the 5 belt system and on the single belt in comparison with the road tests.

different configurations. However, in general, the accuracy of the measurement is lower than for the 5 belt system and could be detrimental to determining the absolute values of drag and lift.

In the case of the drag, for example, there are different interference effects that have to be taken into account, see [13].

At Porsche, the contribution to the drag of the support posts, the restraint bars and the wheel adapters can be easily investigated, by doing a measurement using the 5 belt system. Conversely other forces not belonging to the car, like the friction on the belt and the interference effects between vehicle and the belt itself, are still under investigation.

6 Delta Car-Configurations in Different Wind Tunnels

Because measurements have been carried out in several different wind tunnels in different geographical locations, it is possible to make a comparison similar to the typical "EADE Correlation Test" between different car configurations in these wind tunnels.

As is well known, with the advent of moving ground simulation, the focus of vehicle aerodynamic development has changed in recent years. More attention to detail is necessary in the development of the underbody, as well as for the rims and the rear spoiler.

The car modifications that have been tested in the various wind tunnels visited have been chosen in order to reflect this trend, and are the following:

- Air inlet closed vs. opened (mock-up)
- Rear spoiler extended vs. retracted
- With/without an underbody panel between the rear wheels (in red in Fig. 18)
- With/without front wheel spoilers
- With/without rear wheel spoilers
- Closed rims vs. opened

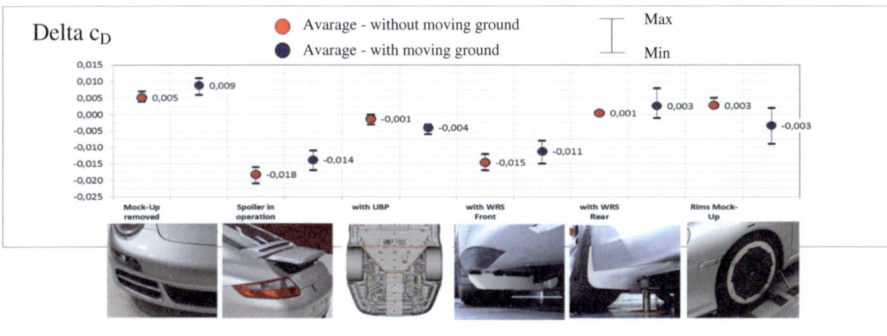

Fig. 18. Delta c_D for different car configurations

Figure 18 shows the trend of the drag delta for these different car modifications in different wind tunnels, including Porsche. The average value measured without moving ground simulation is shown in red, and that with moving ground in blue. The vertical line shows the range between the minimum and the maximum value.

For the sports car reference vehicle, the cooling drag increases with moving ground and the effect of the rear spoiler is altered, producing 4 counts more drag for 10 counts of increased downforce.

In the underbody, the panel between the rear wheels gives a 3 count drag improvement. The importance of the front wheel spoilers is reduced with less drag benefit as expected (11 vs. 15 counts) and 10 counts increase in lift.

The rear wheel spoiler was neutral without moving ground and with moving ground shows a disadvantage. The spread between different wind tunnels is not negligible. The delta swings between −1 and +8 counts. The spread is bigger between the different 5 belt wind tunnels, where the car mounting system and the different gaps between the belts, WRUs and suction systems have an influence on this add-on part. The deltas measured in single belt wind tunnels are conversely almost the same.

Configurations with closed rims also show a discrepancy between different wind tunnels with moving ground.

The spread is from a minimum of −9 counts to a maximum of +2 counts. This range is excessive for an area that has now become much more important due to the WLTP regulations. It was decided to investigate the rim deltas on the road and also in the new Porsche wind tunnel, where it is possible to measure the ventilation drag of the wheels.

7 Ventilation Drag in the New Porsche Wind Tunnel

Ventilation drag is likely to become an increasingly important area of future aerodynamic development.

The wheel rotation creates an aerodynamic torque at the wheel axle. The sources are the surface friction (on tires and rims) and the pressure upstream and downstream of the spokes.

This force, the ventilation drag, is an additional resistance that it is not included in the drag measured by the main wind tunnel balance. It is an internal force between the wheel and the WRU. In some of the new generation wind tunnels, such as the new Porsche wind tunnel, the moving ground system has been improved and additional force transducers have been installed in the WRUs in order to measure this internal force (Fig. 19). A methodology to separate the ventilation drag from the rolling resistance has been developed and published in 2016 [14, 15].

As reported in [14, 15] each wind tunnel measures a different absolute value of ventilation drag, however deltas between different configurations are the same in different wind tunnels.

The same procedure was used in the Porsche wind tunnel to measure the ventilation drag in the standard configuration with and without covered rims.

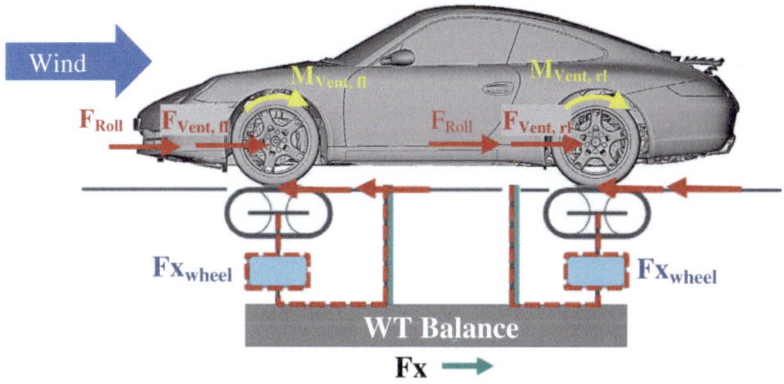

Fig. 19. General sketch of the 5 belt system with additional force transducers, to measure the Fx_{wheel}

The configuration with covered rims showed a reduction of 4 counts in ventilation drag and of 4 counts in the overall car resistance. The sum of these two contributions should be the "real" difference in drag between these two car configurations.

If we assume a reduction of 4 counts in ventilation drag for all the wind tunnels where we measured the configuration with covered rims, the total drag delta effect of covered rims changes to between −13 counts and −2.

Consequently it was decided at Porsche to check this configuration on the road, to have a better understanding of the potential drag reduction.

8 Wind Tunnel vs. Coast-Down

In the calibration phase of the reference cars, several road-tests were performed that focused on the pressure distribution on the vehicle surface. It was important to avoid weather effects and to measure at a constant speed over a time period as long as possible. Coast-down tests were therefore not carried out in the first phase of this project. During the final adjustment of the Porsche wind tunnel, no additional road-tests were carried out, in order to avoid accruing damage to the cars.

When the Porsche wind tunnel was already officially in operation, it was decided to go back on the track with the aim of getting a better understanding of the discrepancy between wind tunnel and coast-down results.

Modern vehicles have different active systems on board, such as cooling air intake shutters, air suspension and spoilers. Sometimes it is difficult to be sure of the position and precise function of these active systems when testing on the road, whereas in the wind tunnel this is relatively well controlled.

Furthermore, during the road-tests changing weather conditions, such as light winds, asphalt temperature and so forth, can have an effect on the results.

The reference car was configured in order to check not only the effects of the local wind (by the Pitot tube and the pressure sensors on the front of the car) but also the tire temperatures and the asphalt temperature. These last two variables have an influence on the rolling resistance and it is important to understand them.

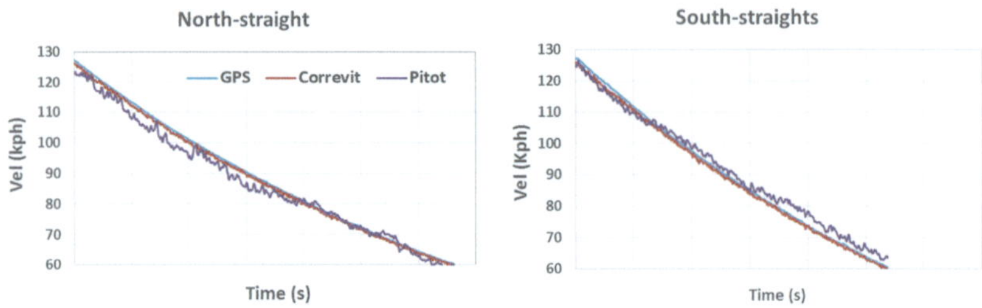

Fig. 20. shows the "driving" Velocity measured from the GPS and from the Correvit and the "real" Velocity measured from the Pitot during a measurement on the North and one on the South Straight.

The data thus gathered from the reference car itself was useful in double checking the weather data measured by the weather stations at the test-track (See Fig. 20). At IDIADA, the proving ground where these tests have been carried out, there is more than one weather station and they are located along the border of the track. However, sometimes the weather station data did not perfectly represent the actual conditions experienced by the car on the track.

The tests to check the drag delta with and without covered rims were carried out on the same day and both configurations where measured over the velocity range from 180 to 5 kph. The coast-down speed range was extended over the standard range (from 120 to 15 kph as defined in [16]) in order to improve the resolution of aerodynamic effects over rolling resistance. Tests were carried out in accordance with the ECE83 procedure. Each configuration was measured 6 times in each driving directions.

Weather conditions were optimal during the tests of both configurations.

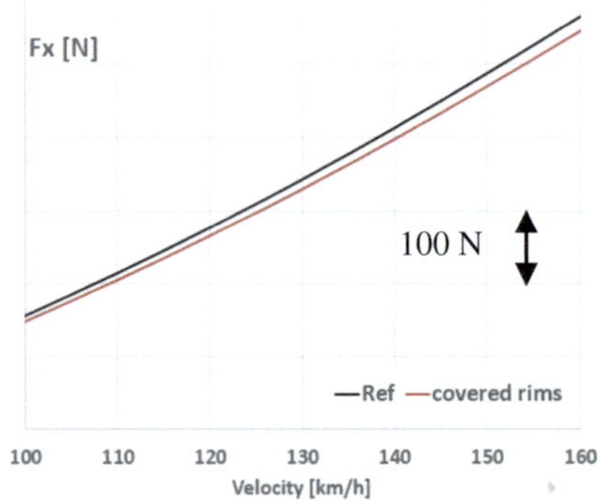

Fig. 21. Coast-down results. Black line: Reference without covered rims. Red line: With covered rims

The results of the coast-down tests show a benefit in drag of approximately 8 counts for the configuration with covered rims (see Fig. 21). This result is in agreement with the results measured in the Porsche wind tunnel (8 counts).

9 Conclusions

Aerodynamics is becoming increasingly more important in the development of modern vehicles, and it is therefore necessary to rigorously validate the tools used in the development process. This is the best way to ensure that there are no surprises when prototype vehicles first take to the road, when the remaining development time is short and modifications difficult and expensive.

A reference car is a powerful aid in developing a better understanding of the aerodynamic differences between conditions on the road and as represented by the standard tools normally used in the development phase.

Comprehensive instrumentation, capable of measuring all relevant parameters with an influence on the aerodynamics, and with an accurate and thorough calibration, are the most important factors in determining the subsequent usefulness of a reference car and value of the results obtained from it.

In particular the opportunity to have two different cars with different blockage effects, and to compare the results measured in several wind tunnels, was important to create a tool with which it is possible to quickly understand a potential problem in a wind tunnel, critical conditions on the road or a critical area in the CFD simulations.

Porsche used these cars for different reasons:

- Firstly, the final tuning of the new wind tunnel. This was a very large and interesting piece of work.
- The goal was to improve the flow stability in the test section, to achieve a pressure gradient on the car surface closer to that measured in the road tests and to improve the repeatability of the aero coefficients.
- Furthermore, the sports car was also used to compare the flow between the climatic wind tunnel and the aeroacoustic wind tunnel. In a climatic wind tunnel, due to the reduced size of the nozzle and the absence of a moving ground, the quality of the flow is unlikely to be good enough to develop and to validate sports cars with air intakes at the back of the vehicle or on the underbody.
- In addition, the reference cars have also been used to compare the pressure distribution and the flow around the vehicles measured in the wind tunnel and simulated by CFD.
- Last but not least, the reference cars were used to check the differences in drag of two car configurations (open and closed rims), when tested in the wind tunnel and by coast down. This part of the work is still in progress and will be expanded in the future, given the emphasis placed on detailed component deltas by the new WLTP regulations.

Acknowledgement. The preparation of the cars, the measurements in all of the various wind tunnels and the final tuning of the new Porsche wind tunnel was a huge task. The authors wish to acknowledge several colleagues at Porsche, who helped in this challenging and interesting project.

Dr. G. Estrada, before leaving Porsche, started this project and his know-how, as shared through his PhD work, has been an important contribution.

Mr. T. Weiser did a huge work in the Workshop to transform a normal car in a precise "reference car".

Mr. S. Gehrmann and Mr. M. Frasch for their support with the hardware and software installation. They were always important to solve last-minute problems.

The Expert Group, Dr. T. Wolf, Dr. R. Mueller and Mr. J. Meder for their useful support and contribution in the final adjustment of the new wind tunnel.

The wind tunnel operators and workers for their efforts during the large number of measurements.

Mr. K. Thomann and Mr. H. Denzinger of the Porsche Energy Management Department for their contribution in conducting the coast-down tests.

A special thank you is also addressed to the colleagues of the external wind tunnels, who gave the authors the possibility to measure our reference cars and to have open and constructive discussions.

Acronyms

CFD	Computational fluid dynamics
WRU	Wheel rotation unit
WT	Wind tunnel
AWT	Aerodynamic wind tunnel
NWT	New Porsche wind tunnel
Count	0.001
c_P	Pressure coefficient
c_p^*	Pressure coefficient measured from the reference car (it means calculated with p_{Ref} instead of p_∞)
p_{ref}	Reference pressure on the bonnet of the car
c_D	Drag coefficient
c_{D_vent}	Ventilation drag coefficient
Fx_{wheel}	Ventilation and rolling resistance on the wheel
F_{Roll}	Rolling resistance
F_{Vent}	Ventilation drag

References

1. Duell, E., Kharazi, A., Muller, S., Ebelking, W., Merker, E.: The BMW AVZ Wind Tunnel Center. SAE 2010-01-0118
2. Stumpf, H., Röser, P., Wiegand, T., Pfäfflin, B., Ocker, J., Müller, R., Eckert, W., Kroß, H.-U., Wallmann, S.: The new aerodynamic and aeroacoustic Wind Tunnel of the Porsche AG. In: FKFS Symposium (2015)
3. Estrada, G.: Das Fahrzeug als aerodynamischer Sensor. Band 54 IVK Stuttgart
4. Estrada, G., Wiedemann, J., Widdecke, N.: The vehicle as an aerodynamic sensor in the Wind Tunnel and on the road. In: FKFS Symposium (2007)
5. Mercker, E., Wiedemann, J.: On the correction of interference effects in open jet wind tunnels. SAE Paper No. 960671 (1996)

6. Mercker, E., Wickern, G., Wiedemann, J.: Contemplation of nozzle blockage on open jet wind tunnels in view of different Q-determination techniques. SAE Paper No. 970136 (1997)
7. Mercker, E., Cooper, K.R.: A two-measurement correction fort the effect of a pressure gradient on automotive, open-jet, wind tunnel measurements. SAE Paper No. 2006-01-0568 (2006)
8. Mercker, E., Cooper, K.R., Fischer, O., Wiedemann, J.: The influence of a horizontal pressure distribution on aerodynamic drag
9. Final Report EADE Correlation Test (2010)
10. Meder, J., Stumpf, H., Röser, P.: Der neue aerodynamische und aeroakustische Windkanal der Porsche AG. VDI Wissensforum, Simulation und Erprobung in der Fahrzeugentwicklung (2016)
11. Hucho, W.H.: Aerodynamik der stumpfen Körper Ed., p. 434, Fig. 9.8 (2011)
12. Collin, C., Mack, S., Indinger, T., Müller, J.: A numerical and Experimental Evaluation of open jet wind tunnel. SAE 2016-01-1597
13. Yano, S.: Drag force measurement accuracy improvement on a full-scale single belt automotive wind tunnel. SATA Meeting (2015)
14. Link, A., Kuthada, T., Widdecke, N., Wittmeier, F., Wiedemann, J.: Analyse, Messung und Optimierung des Ventilationswiderstands von Pkw-Rädern – FAT Schriftenreihe Nr. 291, Verband der Automobilindustrie (VDA), Berlin (2016)
15. Link, A., Kuthada, T., Widdecke, N., Wittmeier, F., Wiedemann, J.: Bestimmung des aerodynamischen Ventilationswiderstands von Pkw-Rädern. Haus der Technik (2016)
16. Regulation no. 83, Revision 5, United Nations (2015). (E/ECE/324/Rev.1/Add.82/Rev.5 – E/ECE/TRANS/505/Rev.1/Add.82/Rev.5)

Methodical Investigation of Vehicle Side Glass Soiling Phenomena

Thomas Landwehr(✉), Timo Kuthada, and Jochen Wiedemann

FKFS, Pfaffenwaldring 12, 70569 Stuttgart, Germany
Thomas.landwehr@fkfs.de

Abstract. Bad weather conditions such as rain and snow affect driving comfort and safety. An unobstructed view onto the surrounding traffic is indispensable. Water whirled up by preceding vehicles can soil the vehicle windows. Depending on the resulting soiling pattern, this can impair the visibility. Therefore, the origin of the vehicle side glass soiling and its influencing factors were investigated in a wind tunnel, resulting in the identification of different soiling phenomena. A crucial parameter is the surface condition as a function of the velocity. Additionally, a new evaluation method is presented which detects and classifies the different types of water accumulations and illustrates the local soiling frequency. It provides information about the composition of soiling patterns and thus can support to during the development process for side glass soiling.

1 Introduction

When driving a car, the view onto the surrounding traffic must be ensured at all times. Especially the view through the side glass onto the rear view mirror and other traffic participants is very important.

While driving on a wetted road, water particles are whirled up by the tires [1]. Some of these droplets are drawn into the wake and generate a spray cloud. Following vehicles get soiled by these small airborne droplets. In addition, the window can also get wetted by direct rainfall.

These surface bound water accumulations can reduce the visibility. They distract the driver especially when they move or do not occur continuously. If a water pool is irradiated by a light beam, a part of the beam will be refracted. Depending on the lighting situation, these refractions can brighten the droplets which can lead to distraction and glare even if one is not looking directly into the light beam [2]. In some situations, it is inevitable to look through the water pool as it is in the view field. Hence, the water pool can obstruct the view [3].

For this reason, soiling tests are usually carried out in a wind tunnel. Thereby, the spray cloud resulting from vehicles driving ahead is simulated by a water spray rake in the nozzle of the wind tunnel. A fluorescence medium is added to the water and this mixture is illuminated with UV light to visualize the soiled areas. Post-processing tools can evaluate the intensity and the degree of soiling [4].

The soiling intensity depends not only on the amount of water impinging the vehicle and its driving speed [5, 6]. Water accumulations are driven by gravity and

J. Wiedemann (ed.), *Progress in Vehicle Aerodynamics and Thermal Management*,
https://doi.org/10.1007/978-3-319-67822-1_16

wind shear forces depending on the inclination angle of the vehicle surface. Windscreen wipers accelerate the carry off. The shape of the A-pillar and the rear view mirror has a decisive influence on the soiling behaviour. The wake of the mirror causes airborne droplets and droplets dripping of the mirror housing to hit the side glass [7]. Last but not least, the surface condition of the side glass has also a crucial impact on the resulting degree of soiling. It influences the wettability and thus the ability of droplets to roll off and how the water accumulates [8]. All this results in droplets and rivulets of varying sizes and shapes arise and might impair the view through the side glass onto the rear view mirror.

Despite the many influencing factors and the resulting different water accumulations on the side glass, the soiling pattern can be attributed to only few phenomena. Therefore, these soiling phenomena were investigated and simulated in the wind tunnel and described below in detail. It is pointed out, which type of water accumulation soils the side glass and where they origin from. The individual phenomena are simulated by emitting water at these origins. Particular attention is paid to the wettability of the side glass and its influence on the soiling as a function of the velocity. In order to evaluate these investigations, the evaluation program DiVeAn was supplemented by additional functions to detect and classify the different types of water accumulations and to illustrate the local soiling frequency.

2 Soiling Process and Evaluation Method

At on-road tests, the existing boundary conditions like velocity and direction of the approach flow, water ingress, road surface condition etc. cannot be reliably influenced. Therefore, soiling tests are usually carried out in the wind tunnel under reproducible conditions.

A defined water volume flow is emitted via spray rack into the air flow to simulate the foreign soiling due to airborne droplets whirled up by vehicles driving ahead. An irrigation device in front of the contact patch area of the tire also allows to simulate the self soiling.

In order to identify the resulting soiling phenomena, the origins of the side glass soiling were identified and simulated by emitting water at these origins. The soiling pattern resulting from these simulations are subsequently compared with the investigations with the spray rack.

The surface condition affects the ability of water to roll off and thus the resulting soiling pattern. Therefore, treatment agents were applied to the side glass producing the desired surface condition.

The soiled areas can be visualized with a camera by adding a fluorescence medium to the water and illuminating the side glass with UV light. Figure 1(a) shows an image resulting from a soiling test with irrigation by the spray rack.

Post-processing tools evaluate the extent of soiling on the side glass surface. The image processing program DiVeAn (Digitale Verschmutzungs-Analyse, engl.: Digital soiling analysis) analyses the grey value of a digital image with respect to the colour intensity of the blue channel of the RGB colour space using an 8-bit discretization. This results in the dimensionless CF value called **C**ontamination **F**actor. This is a

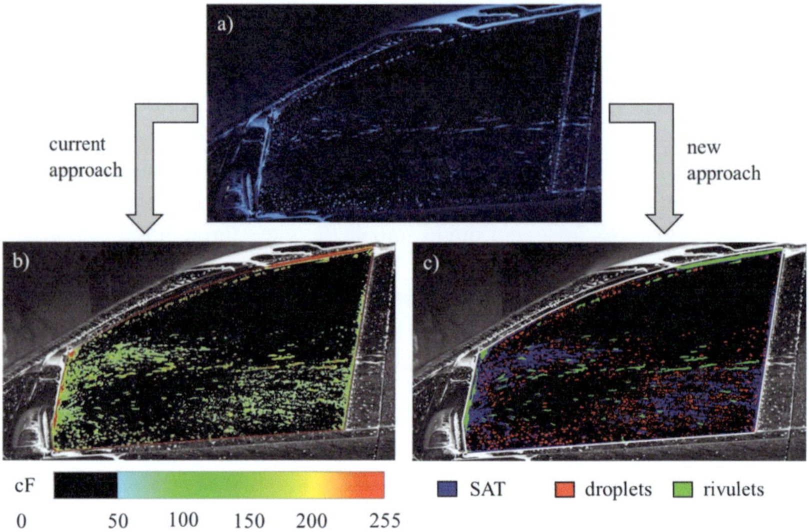

Fig. 1. Evaluation methods of side glass soiling tests (a) Raw image resulting from a soiling test with irrigation by spray rack at v = 110 km/h; (b) Common DiVeAn evaluation detects soiled areas on the side glass; (c) New DiVeAn function distinguishes individual droplets, rivulets and SAT (spray and thin film)

quantitative measure for the emitted light intensity of each pixel within the region of interest and can show values between 0 and 255.

The VG value (Verschmutzungsgrad, engl.: Degree of soiling) indicates the soiled area in percent and is defined as seen in Eq. 1. A threshold value, called FW (Filterwert, engl.: Filter value) defines which light intensity is distinguished as soiled [9]. Areas above this threshold are displayed with their CF value in a miscoloured image, which can be seen in Fig. 1(b).

$$VG = \frac{\sum_{i=1}^{n}(CF_i > FW) \cdot 100}{n}[\%] \tag{1}$$

Depending on the driving condition and weather situation, different types of water accumulations can soil the side glass. For this purpose, DiVeAn has been extended by an additional function for the automatic recognition, classification and evaluation of droplets and rivulets. The new function evaluates the existing grey image output of DiVeAn and distinguishes individual objects with different geometries by means of the Sauvola algorithm. This algorithm, which originates from the binarization of text documents, uses the standard deviation of the grey values to calculate a local threshold value recognizing objects [10]. A support vector machine can classify these objects into categories by machine learning [11]. Objects in a category usually have the same characteristics. Based on these characteristics the types of water accumulations can be classified into single droplets and rivulets. Spray consists of many small droplets, which would be smaller than one pixel and thus too small to resolve even with an 18

Megapixel camera taking a picture of the whole side glass. Therefore, they cannot be distinguished as objects while individual droplets distinguished from the program require a minimum diameter of $d = 1$ mm. It turned out that a droplet diameter of d = 13 mm was not exceeded in all investigations. The reason for this is that small water accumulations are static or roll off slowly and thus they are detected as individual droplets. As the volume of the water accumulation increases, the driving forces from wind and gravity increase too. This results in a faster roll off leading to the formation of rivulets. Depending on the air velocity and the side glass surface condition, more or less droplets are formed. However, there is a similar trend. The most droplets have a droplet diameter around $d = 1$ mm and the number of droplets decreases continuously with increasing diameter. An exemplary droplet size distribution of individual droplets can be seen in Fig. 2.

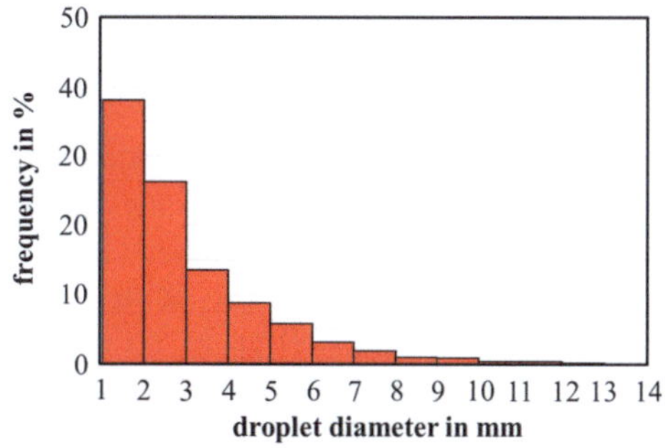

Fig. 2. Droplet size distribution of individual droplets on the side glass

The soiled areas remaining have a low emitted light intensity and a small light intensity gradient and can be classified as spray and thin film areas, hereafter called SAT. This combined with the detection of rivulets and droplets enables the classification into areas soiled by droplets, rivulets and SAT as can be seen in Fig. 1(c).

Figure 3 shows the VG of the test displayed in Fig. 1. DiVeAn detects the total soiled area and the new added function distinguishes the individual droplets, rivulets and SAT within this area.

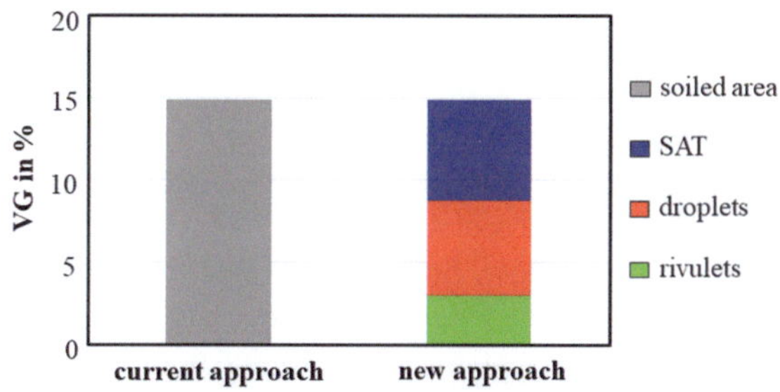

Fig. 3. The new function in DiVeAn enables the classification of the soiled area into different types of soiling

3 Phenomena of Vehicle Side Glass Soiling

The flow field around the vehicle and thus the soiling is mainly affected by the shape of the vehicle. The soiling of the side glass depends on various factors. The driving speed, as well as the amount and direction of the water impinging are factors which are independent of the vehicle and cannot be influenced during the vehicle development. However, the vehicle itself has a decisive influence. The vehicle shape affects the airflow and the surface condition is crucial for the ability of water to roll off. The use of channels, steps and seals can influence the flow behaviour. Despite the many different types of water accumulations that can soil the side glass, the soiling can be traced back to only a few origins.

The rain soils the side glass especially at low velocities resulting in droplets distributed over the entire surface by direct raindrop impact [6]. The side glass can also get soiled by the splash of vehicles passing by. Since these types of soiling cannot be influenced by changes of the vehicle geometry, they are usually not investigated in wind tunnel tests.

Rotating wheels cause turbulence due to their rotation. For passenger cars and SUVs, this rotation does not exert any influence on the side glass soiling. The self soiling can be simulated by placing an irrigation device in front of the contact area between roller and tire. As a result, droplets get swirled up and soil the vehicle. However, this usually only soils the side wall up to the door handle, while the side glass remains free from water accumulations. In the case of sport cars, the soiled area could also reach the lower part of the side glass.

The main source of the side glass soiling is the spray of preceding vehicles. This generates a particle-laden flow that can soil the side glass. The resulting soiling is caused by a few phenomena which occur depending on the vehicle shape and the velocity of the oncoming flow as seen in Fig. 4.

The crucial aerodynamic influencing factors are the A-pillar overflow and the flow around the exterior rear view mirror [12]. These soiling phenomena are subsequently explained and simulated by emitting water at their origins (Figs. 5 and 6).

A part of the particle-laden flow caused by the spray of preceding vehicles impinges the windscreen. Gravity and wind driven shear forces act on the surface bound water accumulation which are carried away depending on the inclination angle of the wind screen and the surface condition. An equilibrium of forces leads to the formation of a water pool. The function of the wiper is to keep the windscreen free of water. In this process, the wiper transports the water towards to the A-pillar. The curvature of the A-pillar induces a vortex with a negative pressure area, which acts on the water accumulation. At low velocities, this has no effect on the side glass soiling (Fig. 4(a)). With higher velocities, the airflow over the A-pillar increases also whereby a water overflow onto the side glass can occur (Fig. 4(b) and (c)).

The wiper supports this overflow. Studies show that the wiper leads to a periodic overflow [13]. Large steps between the front window and the A-pillar or water management channels reduce or prevent the overflow by draining the water. However, these geometrical measures can lead to acoustic and aerodynamic disadvantages. Therefore, the development process is always a compromise between different

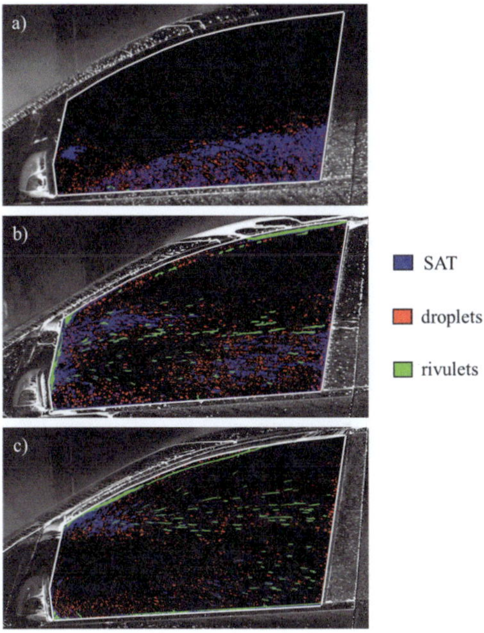

Fig. 4. Soiling phenomena on the side glass caused by irrigation with a spray rack (a) Spray at the front region and soiling wedge at v = 50 km/h due to the circulation of the exterior rear view mirror (b) Horizontal rivulets at v = 110 km/h due to A-pillar overflow (c) Rivulets along the A-pillar due to A-pillar overflow at v = 150 km/h

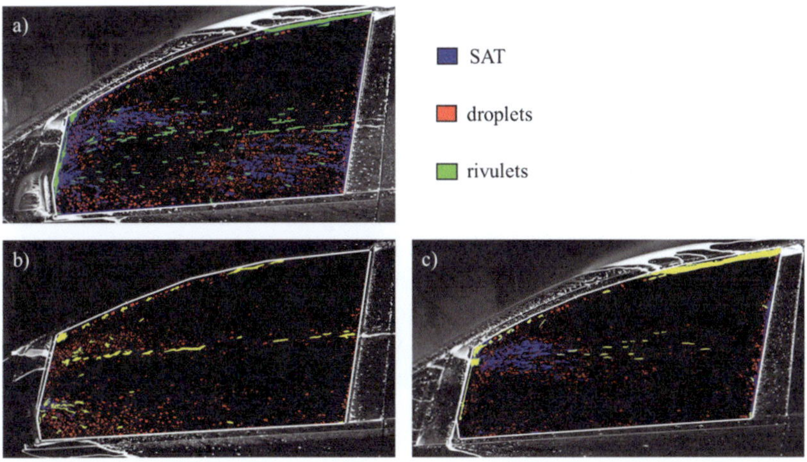

Fig. 5. Simulation of the side glass soiling at v = 110 km/h resulting from A-pillar overflow (a) Irrigation by spray rack, vehicle with exterior rear view mirror (b) Irrigation by emitting water at the mirror triangle, vehicle with rear view mirror (c) Irrigation by spray rack, vehicle without rear view mirror

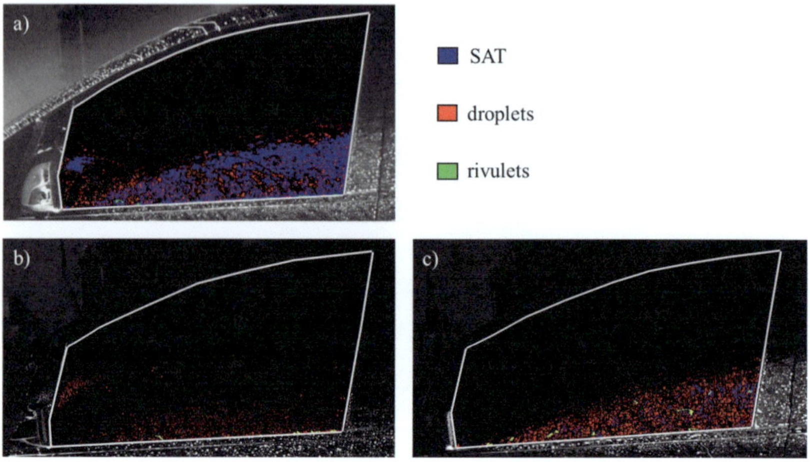

Fig. 6. Simulation of the side glass soiling at v = 50 km/h induced by the rear view mirror (a) Irrigation by spray rack (b) Irrigation by emitting water 15 cm in front of the gap between the mirror housing and the vehicle door (c) Irrigation by emitting water at the dripping point of the rear view mirror

disciplines. For design reasons, the height of the water management channels decreases downstream. This favours a transverse flow and an overflow onto the side glass.

The majority of the water which leads to the A-pillar overflow arises due to direct droplet impact on the A-pillar. Hence, the surface should be kept as small as possible to reduce the amount of droplet impact. The impinging water as well as the transverse flow gets carried downstream due to the high shear forces imparted by the airflow.

Obstacles such as trims or margin seals can influence the overflow onto the side glass. The water is transported downstream in the gap due to gravity and capillary forces as well as the less pronounced shear forces in the trailing area and emerges at the mirror triangle.

It is possible to simulate the A-pillar overflow by a selectively emitting water at this point as shown in Fig. 5(b). The comparison with the image in Fig. 5(a) as a result of foreign soiling shows some similarities. In both cases, single droplets and rivulets are moving horizontally across the side glass and along the A-pillar. In addition, in the area behind the exterior rear view mirror many individual droplets can be seen. The spray and droplets in the rear lower part of the side glass, commonly referred to as soiling wedge, did not form during the investigations for simulating the A-pillar overflow.

The exterior rear view mirror induces a wake affecting the airflow by its shape, size and the position of its mounting. Due to this wake, additional forces act on the surface bound water accumulation and change their path. On some vehicles, this can lead to a low pressure region on the side glass resulting in an additional water pool at mirror height releasing droplets and rivulets. In order to examine the influence of the presence of the entire mirror geometry on the side glass soiling, the mirror with its mount was removed and replaced by a flat cover (Fig. 5(c)). Here, as well as in the investigations including the mirror shown in Fig. 5(a), the water overflow from the windscreen and the resulting rivulet appear at the rear upper end of the side glass. The spray which can be detected is formed in the upper front part of the side glass. The spray and droplets in the rear lower part of the side glass cannot be identified in the tests with the simulated A-pillar overflow or without the rear view mirror. Therefore, these droplets have to origin from the mirror and its mounting. This study shows that the rear view mirror wake does not influence the A-pillar overflow.

However, the resulting soiling pattern due to A-pillar overflow depends also on the vehicle speed respectively the wind velocity. At low velocities, the gravity is predominant and the water flows downwards on the side glass along the edge. The A-pillar vortex induces an upward-directed force component. In the case of force equilibrium, the water accumulates and forms a pool. Constantly added mass by incoming water and the forces applied by the airflow induce an imbalance which leads to varying pool size and position. The equilibrium is maintained by emitting droplets and rivulets moving across the side glass, passing the view field of the driver (Fig. 4(b)) onto the rear view mirror. Increasing velocity enhances the A-pillar vortex resulting in rivulets moving upwards along the A-pillar, which is perceived as less critical (Fig. 4(c)).

When the airborne droplets hit the mirror housing, they can splash into many small droplets and get entrained into the airflow. The splashing depends on the impact energy and the counter-acting surface tension of the droplet. The wake of the mirror causes airborne droplets to hit the side glass. At low velocities, this results in a spray in the front region of the side glass covering large parts of the view field. These phenomena cannot be prevented even by means of a concerted design of the gap between the mirror housing and the vehicle or a modification of the mirror shape. It was possible to simulate this soiling phenomena by emitting water 15 cm in front of the gap between the mirror housing and the vehicle door. However, the simulation of this phenomenon led to larger droplets while the investigations with the spray rack results in spray (Fig. 6(b)).

A part of the droplets impinging the mirror surface generate a water film, which is being driven downstream by the wind. As it reached the mirror edge, it drips off. Entrained by the airflow, the droplets soil the vehicle forming the characteristic soiling wedge as can be seen in Fig. 6(a). This phenomenon is more pronounced at lower velocities.

Since the rear view mirror was not soiled in the investigations of Fig. 5(b) and (c), this soiling wedge is not formed either. To reproduce the soiling by mirror dripping droplets, the mirror geometry was obtained by means of a 3D scanner, modified and subsequently built by a rapid prototyping machine. This enables a defined water outlet at several positions at the mirror edge and mirror surface generating the soiling wedge as shown in Fig. 6(c). Again, the spray could not be resolved as finely as in the reference measurement.

The location of the dripping point at the rear view mirror was mostly independent of the position of the water outlet at the mirror. However, it can be affected by water guiding features like grooves, lips or contour changes. These features are designed to prevent dripping droplets from getting into the wake or hitting the side glass further downstream.

In addition to the side glass, the mirror glass must be kept also free of water accumulations in order to allow the driver to view the rearward traffic. Here, the mirror wake leads to a recirculation resulting in airborne droplets hitting the mirror glass. This phenomenon is mainly fed by droplets dripping of the mirror housing or by droplets from the oncoming flow. The influence of the A-pillar overflow on the mirror glass soiling is negligible. The irrigation by emitting water at the mirror triangle led only to a few isolated droplets on the mirror glass at higher velocities.

4 Influence of Contact Angle on Vehicle Side Glass Soiling

The surface condition affects the droplet roll off and thus the soiling intensity. The wettability is defined by the contact angle and describes to which extent a liquid spreads when it comes into contact with a solid. The contact angle is dependent on the surface free energy of the solid σ_l, the surface tension of the liquid σ_s and the interfacial tension between liquid and solid σ_{sl} (Fig. 7) and can be determined by the integration of the Young's Equation (Eq. 1) [14].

$$\sigma_l \cdot \cos\theta = \sigma_s - \sigma_{sl} \tag{2}$$

Realistic contact angles on the side glass are $\theta = 30° - 75°$. The composition of the glass plane has an influence on the contact angle. A polished side glass features usually a contact angle below $\theta \sim 25°$. Treating agents increasing the contact angle are applied to the surface when washing the vehicle in the car wash. Different wash programs from several providers produce a contact angle of $\theta = 55° - 70°$. Over the time, the contact angle is reduced due to deterioration of the treatment agent. Special coatings can lead to a higher contact angle which a common car wash is not able to reach. By applying different treatment agents, it is possible to chance the surface condition of the side glass to create the desired contact angle.

The influence of the contact angle on the side glass soiling at v = 110 km/h is displayed in Fig. 8. Figure 9 shows the average VG of the side glass depending on the contact angle and the velocity for the last 30 s of each run. Contact angles $\theta < 20$ supress the formation of droplets and rivulets which results in a thin spread water film. Thus, much higher degrees of soiling occur over the entire considered velocity range.

Lager contact angle enable the formation of droplets and rivulets. The contact area between the water and the glass plane is reduced and thus the film height is increased. This increases also the cross sectional area on which the airflow acts. Together with the reduced contact area, this enhances the roll off behaviour. Thus the side glass soiling, especially at high velocities, is decreased.

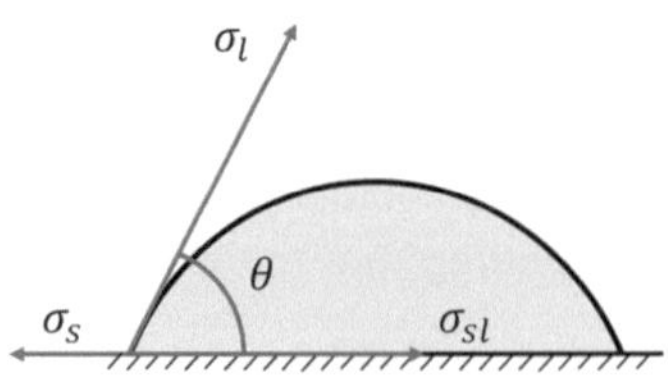

Fig. 7. Contact angle of a schematic droplet on a solid surface

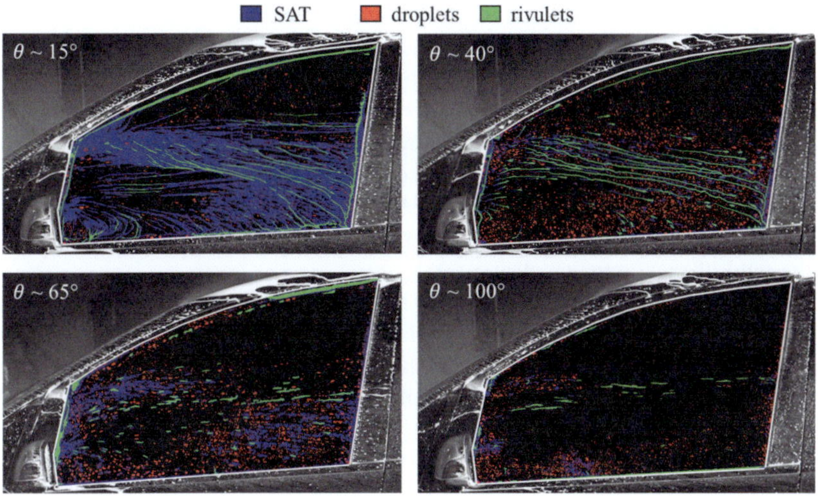

Fig. 8. Influence of contact angle on the side glass soiling at v = 110 km/h

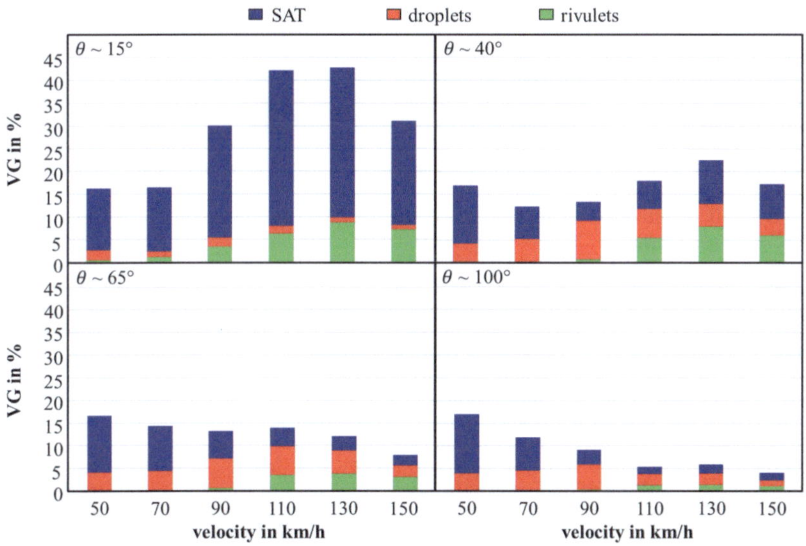

Fig. 9. VG of the side glass depending on the contact angle and the velocity

Contact angles $\theta > 30$ show a comparable soiling behaviour on the side glass at low velocities. Here, the soiling wedge consisting of individual static droplets and spray is formed. In addition, the spray at the front region of the side glass is build (Fig. 4(a)). The portion of the rivulets at the VG is negligible. As the velocity increases, the driving force acting on the water accumulations results in the formation of rivulets and a decreasing portion of SAT. Starting with v = 90 km/h, a noticeable influence of the contact angle on the side glass soiling can be observed.

Long rivulets are formed for $\theta \sim 40°$, which are moving across the side glass through the field of view. In this case, a share of SAT results from residues of rivulets which have already moved downstream. For higher velocities v = 110 km/h, the build up of the soiling wedge is prevented. At higher contact angles, smaller rivulets and single droplets are formed and roll off faster. At $\theta \sim 100°$, the water accumulations rolls off quickly resulting in an almost unsoiled side glass.

As the velocity increases further, the A-pillar vortex becomes dominant and the rivulets follow along the A-pillar (Fig. 4(c)). This reduces the VG for all contact angles and the field of view remains clear.

The results shown in Fig. 8 are only snapshots. The soiling behaviour changes during the course of each test run and should be considered therefore with respect to its transient development [13]. During the initial phase of each run, the VG increases reaching a quasi-steady phase after some time as displayed in Fig. 10.

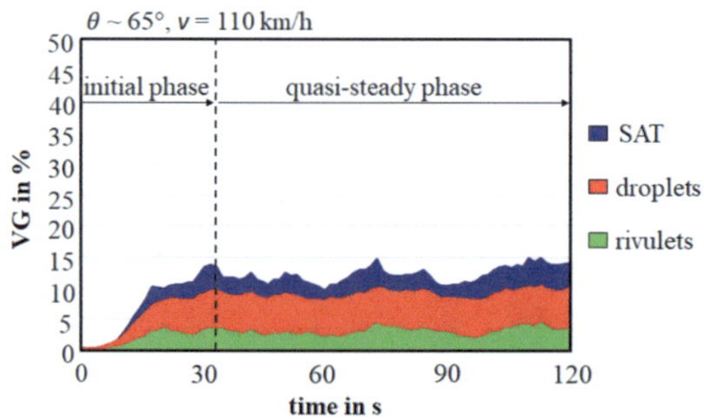

Fig. 10. Exemplary evolution of the VG on the side glass during one test run

After reaching this quasi-steady phase, the accumulation and draining of water are in equilibrium. However, the soiling behaviour is still fluctuating. The A-pillar overflow caused by the wiper can lead to a periodic increase in the VG. This increase can be particularly noticeable in the evaluation of subareas such as the driver's field of view onto the rear view mirror.

Figure 11 shows the standard deviation of the results from Fig. 9. Droplets have the smallest portion of the fluctuations. Higher velocities and higher contact angles do not only lead to a lower VG of soiling, but also to a lower standard deviation. At low velocities, SAT has a higher portion of the VG and large fluctuations.

Since the evaluation of only one time step can lead to misinterpretations, Vollmer [8] proposed the visual representation of a transient evaluation representing the frequency of the local soiling in a certain time interval. Figure 12 shows this illustration for the test already presented in Fig. 1. It depicts areas which are soiled more often and areas which remain clear. This evaluation is particularly useful when tests feature fluctuating soiling pattern. However, it does not provide information on the time evolution of the soiling or the flow direction. Therefore the results of each time step should be considered.

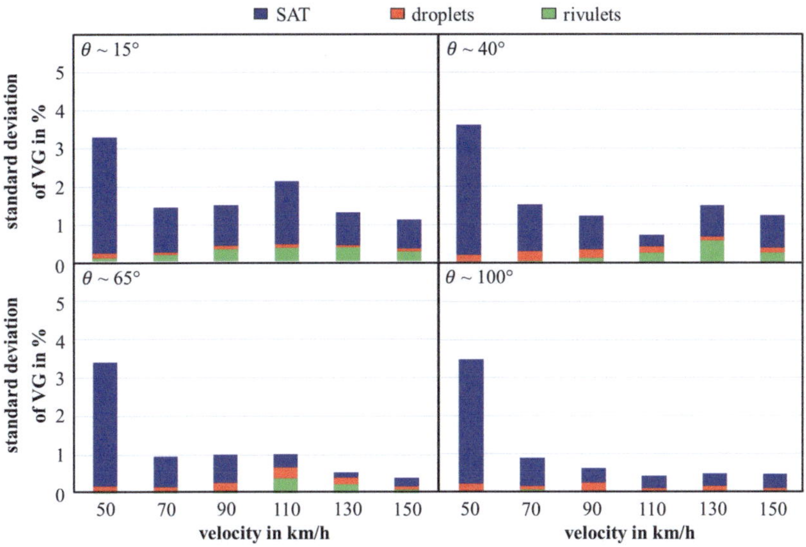

Fig. 11. Standard deviation of VG for the last 30 s resulting of the evaluation from Fig. 9

Each evaluation method has its advantage and hence its eligibility. The current evaluation DiVeAn method provides information on the film height and the soiled area of chosen relevant areas such as the field of view. The improved approach in DiVeAn enables to classify the soiled area into the types of water accumulation, providing a better understanding of the composition of the soiling patterns and resulting soiling phenomena. The frequency evaluation of the local soiling shows pictographic the areas with a high temporal exposure to water accumulations. The transient course of the VG can support the identification of the quasi-steady phase and the quantification of the fluctuations. Averaging the time steps within the last 30 s of each investigation extenuates the influence of aberrations and fluctuations and therefore increases the reproducibility.

The average VG of soiling is an appropriate measure to compare two different configurations with each other. However, it is not sufficient to evaluate just one

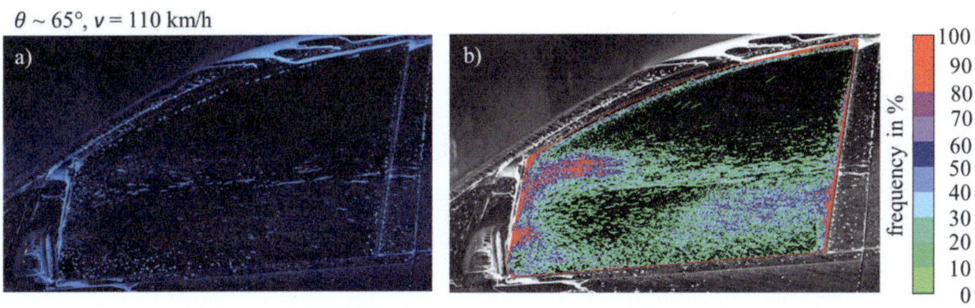

Fig. 12. Visual evaluation of the local soiling frequency (a) Raw image of the last time step (b) Frequency image resulting from the time interval of the last 30 s

characteristic value. A meaningful conclusion about the soiling behaviour of the vehicle and its consequences for the driver can only be drawn from the combination of the introduced visual evaluations.

5 Conclusion and Outlook

The presented study discusses the soiling phenomena of vehicle side glass. Initially, the causes and origins of the vehicle side glass soiling are discussed. The main influence is the foreign soiling caused by whirled up droplets of vehicles driving ahead. These droplets can impair the view through the glass pane of vehicles following in various ways.

To identify the phenomena responsible for the soiling of the vehicle side glass, investigations were carried out in the Thermal Wind Tunnel of FKFS. In order to evaluate these investigations, the FKFS developed program DiVeAn was supplemented by an additional function. This new function enables the detection and classification of water accumulations into individual droplets, rivulets, spray and thin film areas.

The resulting characteristic soiling pattern can be attributed to only a few phenomena. Subsequently, these phenomena are identified, described and individual simulated. Low velocities lead to the formation of the characteristic soiling wedge consisting of spray and droplets originating from the mirror housing. Additionally, spray is formed at the front region of the side glass covering large parts of the view field resulting from airborne droplets out of the mirror wake. As the velocity increases, the influence of the mirror decreases and an A-pillar overflow appears. The water from this overflow is transported downstream on the side glass.

Additionally, the influence of the surface condition in combination with air velocity on the side glass soiling is considered. Contact angles of $\theta < 20$ supresses the formation of droplets and rivulets resulting a thin water film and thus a high degree of soiling. A high contact angle facilitates the ability of water to roll off and reduces fluctuations. This ensures a lower degree of soiling in particular at high velocities. Finally, an evaluation method is presented to illustrate the frequency of the local soiling in a time interval. This supports a visual transient evaluation of the resulting soiling.

These results improve the understanding of the side glass soiling phenomena and their influencing factors. This knowledge can help in the vehicle development to achieve an unobstructed view onto the surrounding traffic and thus to an improved driving safety.

Abbreviations

UV	Ultraviolet
DiVeAn	Digitale Verschmutzungs-Analyse, engl.: Digital soiling analysis
RGB	Red–green–blue
CF	Contamination Factor
SAT	Spray and thin film
VG	Verschmutzungsgrad, engl.: Degree of soiling
FW	Filterwert, engl.: Filter value

References

1. Spruss, I.: Ein Beitrag zur Untersuchung der Kraftfahrzeugverschmutzung in Experiment und Simulation. Dissertation, Universität Stuttgart, Stuttgart (2015)
2. Landwehr, T., Kuthada, T., Wiedemann, J.: Refraction properties of light at wetted vehicle glass planes. In: 17th Stuttgart International Symposium (2017)
3. Landwehr, T., Kuthada, T., Widdecke, N., Wiedemann, J.: Investigation of visibility properties through wetted glass planes on vehicles. In: 16th Stuttgart International Symposium (2016)
4. Widdecke, N., Kuthada, T., Wiedemann, J.: Moderne Verfahrensweisen zur Untersuchung der Fahrzeugverschmutzung. In: Tagung "Aerodynamik des Kraftfahrzeugs", Haus der Technik e.V. München (2001)
5. Karbon, K.J., Longman, S.E.: Automobile exterior water flow analysis using CFD and wind tunnel visualization. SAE technical paper 980035, Society of Automotive Engineers, SAE, Warrendale, PA (1998)
6. Vollmer, H., Gau, H., Klußmann, S., Kuthada, T., Wiedemann, J.: Comparison of on-road and wind tunnel testing of side window soiling using a new method. In: Wiedemann, J. (Hrsg.) Progress in Vehicle Aerodynamics and Thermal Management: Proceedings of the 10th FKFS Conference, pp. S. 276–288. Expert-Verlag, Renningen (2015)
7. Bannister, M.: Drag and dirt deposition mechanisms of external rear view mirrors and techniques used for optimization. In: Society of Automotive Engineers, Inc. 2000-01-0486
8. Vollmer, H., Gau, H., Winkelmann, H., Kuthada, T., Wiedemann, J.: Methode zur Bewertung der Sichtfreihaltung bei Regen. In: Tagung "Aerodynamik des Kraftfahrzeugs", Haus der Technik e.V. München (2014)
9. Kuthada, T., Widdecke, N., Wiedemann, J.: Advanced investigation methods on vehicle soiling. In: 4th MIRA International Vehicle Aerodynamics Conference, Warwick (2002)
10. Sauvola, J., Pietikäinen, M.: Adaptive document image binarization. Pattern Recogn. **33**, 225–236 (2000)
11. Burges, C.J.C.: A tutorial on support vector machines for pattern recognition. Data Min. Knowl. Discov. **2**(2), 121–167 (1998)
12. Höfer, P., Mößner, A.: Schmutzfreihaltung. In: Schütz, T. (Hrsg.) Hucho – Aerodynamik des Automobils: 6. Auflage, pp. S. 464–484. Springer Vieweg, Wiesbaden (2013). ISBN 978-3-8348-2316-8
13. Spruss, I., Landwehr, T., Kuthada, T., Wiedemann, J.: Advanced investigation methods on side glass soiling. In: Wiedemann, J. (Hrsg.) Progress in Vehicle Aerodynamics and Thermal Management: Proceedings of the 9th FKFS Conference, pp. S. 167–181. Expert-Verlag, Renningen (2013)
14. DIN 55660-1: Beschichtungsstoffe – Benetzbarkeit – Teil 1: Begriffe und allgemeine Grundlagen. Beuth Verlag, Berlin (2011)

Design and First Test of the New Synchronous 200 Hz System for Unsteady Pressure Field Measurement

Jakub Filipský[1(✉)], Jan Čížek[1], Felix Wittmeier[2], Timo Kuthada[2], and Simon Meier[2]

[1] 4Jtech s.r.o., Ringhofferova 115/1, 155 21 Prague 5, Czech Republic
jakub.filipsky@4jtech.cz
[2] FKFS, Stuttgart, Germany

Abstract. This paper introduces an innovative pressure measurement system for unsteady flow measurement. Unsteady flow is currently one of the hot topics of road vehicles aerodynamics and an inexpensive, simple-to-use tool for its investigation has long been missing. Conventional pressure measurement systems with centralized pressure scanners and long tubings do not satisfy the needs of unsteady phenomena. An approach of miniature MEMS pressure sensors is used instead, whereby the sensors are installed directly on the surface of the body under investigation. This concept not only enables the possibility of unsteady flow measurement but also offers a fast and reliable method of pressure field investigation. Finally, the measurement system was aimed at the specific needs of road vehicle productive testing and provides an effective tool capable of rapid experiment preparation, reliable measurement and simple post-processing.

The newly developed system was tested and validated in different experiments in a specialized test stand and in model scale wind tunnel measurements.

1 Concept

The system consists of miniature MEMS pressure sensors installed on a flexible Strip which is both part of the mechanical construction of the device and also a Printed Circuit Board (PCB) connecting all electrical signals (Fig. 1). Extremely low overall dimensions are achieved with a maximum height of 1.25 mm. The benefit of placing pressure sensors directly on the place where pressure measurement is needed brings a far higher frequency range compared to the traditional approach of long tubes, which reduce signal propagation at higher frequencies. Replacing pneumatic signals in tubes by electrical signals in wires also increases the system's overall reliability as wires are generally less sensitive to defects than tubes. The all-digital solution of pressure measurement further lowers the risk of misplacing a channel during post-processing.

J. Wiedemann (ed.), *Progress in Vehicle Aerodynamics and Thermal Management*,
https://doi.org/10.1007/978-3-319-67822-1_17

Fig. 1. A 200 Hz Strip with 24 absolute MEMS pressure sensors

2 Characteristics

The development of the measurement system was carried out with several goals in mind:

- Up to 200 Hz sampling frequency
- 24 pressure sensors on a single Strip
- Small outline dimensions
- Up to 20 Strips on a single communication bus
- Expansion of communication bus via multiple synchronized Central Units
- Readings from all sensors (all Strips, all CUs) synchronous within ±1 sample

All goals stated above were successfully met although the development team has faced some difficulties along the way. The following chapter will give details on the individual parts of the measurement system and its properties.

2.1 Topology

The measurement system is built around a digital bus linear topology network with a single master (Central Unit) and multiple slaves (Strips or any other general sensor). The industrial bus RS485 was chosen for the hardware layer of communication as transfer rates exceeding 6 Mbit/second are required. A custom-designed protocol was implemented in higher communication levels to meet the strict requirements placed on the system. There is a great benefit to the concept residing in the single bus communication principle. The bus can be split multiple times and the installation of such a system becomes extremely fast and easy; an example of such a connection is displayed in Fig. 2. Experiments with hundreds of pressure intakes can be prepared in a matter of hours, which means an extreme time reduction compared to standard pressure measurement systems. This benefit is greatly appreciated in productive testing when time

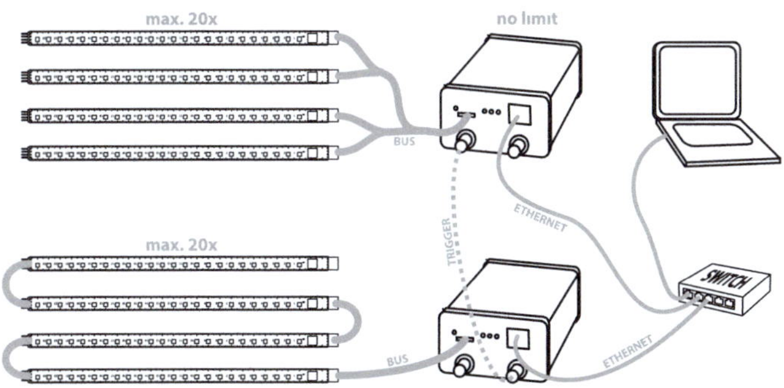

Fig. 2. Connection topology and synchronization of the measurement system

for preparation is often critical and complex experiments could not be carried out otherwise.

2.2 Strips

Each Strip includes 24 pressure sensors 10 mm apart and some supporting circuitry (MCU, voltage stabilizers, indicating LED, and bus driver). The MCU communicates with all onboard sensors and transfers their readings via bus. The readings from all 24 sensors are not perfectly synchronous; some misalignment of ± 0.3 sample occurs. Some other functions are supported by the MCU besides data readout such as memory

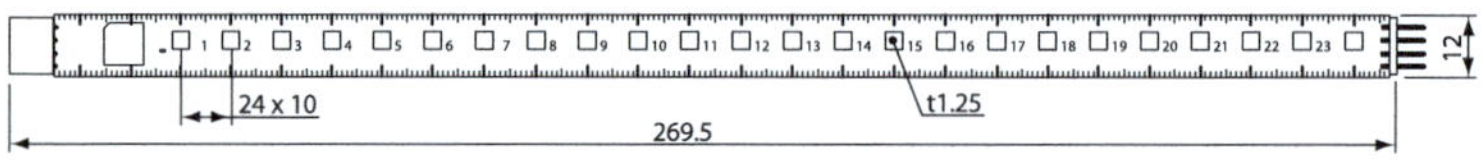

Fig. 3. Dimensions of a Strip in mm, maximum total thickness is 1.25 mm

readout or Strip ID initialization.

The outline dimensions of the Strips are sketched on Fig. 3; the total height of the device is 1.25 mm. Invasiveness of such a device is negligible in the application of road vehicle aerodynamics based on previous experience [1].

2.3 Central Unit

The Central Unit (CU) is a non-transparent logic converter from RS485 (common bus) to the Ethernet connecting the CU to a host computer. The CU synchronizes all Strips connected to the bus, autonomously handles the lowest levels of communication and is capable of searching for newly connected sensors to enable the plug-and-measure functionality. Each CU is equipped with a trigger input and a trigger output (signal generator). The trigger input may or may not be used to trigger the samples when synchronization with other measurement systems is necessary. Alternatively several CUs can be connected together to get a fully synchronized expandable measurement system with virtually no limit on the number of connected sensors.

The CU is supplied by a 12 VDC power supply and can thus be easily used for on-road measurements when battery operation is required (Fig. 4).

Fig. 4. Central unit for communication with host PC

2.4 Sensors

Traditional methods of measuring surface pressures in external aerodynamics suppose the use of differential pressure sensors which measure the pressure difference between a tap on the surface and a reference pressure. When solving the task of scaling the system down and placing the sensors themselves on the surface there would need to be some kind of mechanical solution of how to apply the reference pressure on the bottom side of the pressure sensing elements. This approach would be fairly impractical and potentially unreliable. Absolute pressure sensors were chosen to omit this difficulty. The result of the experiment is the same as in the traditional approach but the difference between surface pressure and reference pressure is not done by the sensor itself (differential measurement) but rather mathematically in the post-processing. Some degree of accuracy is sacrificed for the sake of simplicity.

The sensors of choice are absolute miniature MEMS pressure sensors with an integrated ASIC. The pressure reading is burdened by a strong temperature dependency, which is normal for MEMS pressure sensors. Compensation is done numerically through an integrated temperature sensor.

The sensors themselves operate on a pressure range of 30 to 120 kPa, which is inefficiently wide for use in automotive aerodynamics. To get favorable accuracy two steps were adopted:

Ad-hoc calibration in a narrower pressure range.

Reference pressure is sampled by a dedicated Strip.

Each pressure sensor of each Strip is calibrated with respect to the steps mentioned above. The calibration takes place for varying pressure and varying temperature (normally 93 to 107 kPa and 15 to 40°C) resulting in a calibration surface. The surface is then interpolated by a polynomial fit. Nevertheless, even after the calibration there is some remaining temperature and time drift. The latter adopted step – dedicated Strip for reference pressure – cancels out most of the remaining drifting. Since the drifts introduce a systematic error we can use this property to get a better overall accuracy by subtracting the reference pressure (absolute pressure value) from the surface pressure (also absolute pressure value). Both of these pressures are burdened by a similar drift error and the subtraction cancels it out.

Overall performance of the Strips is summarized in Table 1.

Table 1. Summary of properties and accuracy

Range	
Pressure	93 to 107 kPa
Temperature	15 to 40°C (up to 85°C possible)
Accuracy	
Overall accuracy	±60 Pa
Pressure difference accuracy	±20 Pa
Temperature	±1°C
Noise (3σ values)	
@160 Hz	±15 Pa p2p
@200 Hz	±23 Pa p2p

3 Measurement System

Road vehicle productive testing is driven by the enormous pressure of time-to-market constraints. This often limits the use of sophisticated experimental methods on a productive basis as the time for experiment preparation is counted in hours, not in days or weeks as is usual for laboratory experiments. There is also often no chance to repeat the experiment if any data is not valid or lost in the process. Four specifics targeted by the development of the measurement system were identified:

- A useful experiment needs to be prepared within several hours
- Any fault of the measurement system needs to be reported to the user instantaneously
- The three phases of an experiment (preparation, measurement, post-processing) need to be easily handled by three different people with no unnecessary overhead
- All available information about the measurement system needs to be kept together throughout all three phases of experiment

The physical installation of the Strips on the surface of an investigated body is done by adhesive stickers and is extremely fast and simple. Wiring the devices to the communication bus is not a long process in itself but there is a drawback in the simplicity of a single communication bus. When there is a hardware error in the wiring (misplaced connectors, faulty wire, etc.), the entire system stops working and debugging is fairly troublesome. The chosen solution implements a true plug-and-measure functionality when any newly connected sensors (Strips) are instantaneously displayed to the user and ready to measure. On the other hand, when there is a hardware error the user is notified of which part of the measurement system contains an error and can easily debug the system.

Another innovative functionality of the measurement system makes use of the internal memory of the Strips and of the Central Units. All information necessary for the measurement is stored within the devices themselves and is read whenever a new device is plugged in. Calibration coefficients of sensors are carried inside each Strip; sampling frequency, trigger setting, IP address etc. are stored inside each CU. Every Strip can additionally be issued a user-definable name during installation, which is also stored within the Strip itself. The user-definable layout of the GUI dashboard, position and type of each graph is also stored within the devices and is updated each time a new sensor is found. This concept is different from a traditional measurement system when project files need to be transferred along with hardware when migrating the experimental setup from one person to another. This not only makes teamwork on an experiment very easy process-wise but also makes sure that no information is mismatched when transferring the experiment through phases and through people.

The GUI (see Fig. 5) itself displays the topology of connected sensors, user notifications and a dashboard. The user can interactively set up the dashboard, and change the graph type, graph position, and displayed sensors with their quantities.

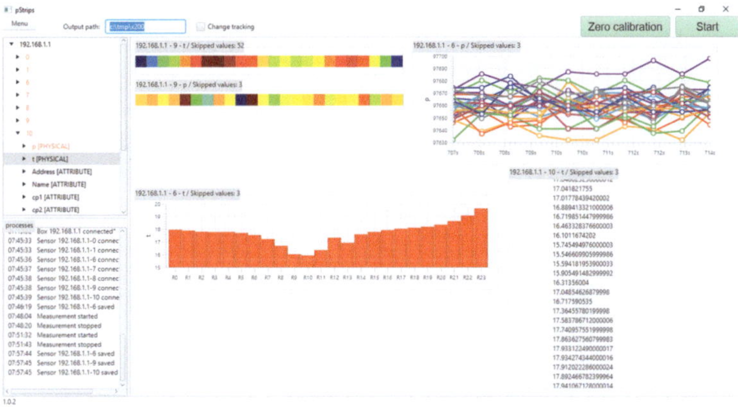

Fig. 5. Screenshot of the graphical user interface for experiment

4 System Validation

After completing the development of the system, the pressure strips were tested at FKFS both in a specialized dynamic pressure test stand and in a model scale wind tunnel experiment.

4.1 Experimental Setup

To validate the dynamic capabilities of the new pressure strip system, a specialized test stand was designed, capable of generating fluctuating pressures in different frequencies with a high power speaker.

In addition to the test stand validation, experiments in the model scale wind tunnel of University of Stuttgart were conducted. Although the pressure strips generally are better suited for full-scale tests, the model scale tunnel was chosen for the validation measurements due to its better availability and to show, whether the size of the pressure strips is small enough to use it also for model scale measurements.

4.1.1 Dynamic Pressure Test Stand

The dynamic pressure test stand consists of a signal generator, an amplifier and high power speaker, capable of generating pressure fluctuations in the magnitude of around 40 Pa in a frequency range of 1–20 Hz. The volume behind the speaker is sealed and the pressure strips are installed in this volume. In addition to the pressure strips, eight pressure tabs were connected to a PSI initium system and eight to Kulite sensors, to measure the reference signals in the sealed volume.

To compare the results, only eight of the channels of the pressure strip were used, allowing direct comparison of the pressures at the same positions relative to the speaker. For the validation, both the averaged value over all eight sensors and the absolute value of each single sensor was evaluated.

4.1.2 The Model Scale Wind Tunnel of University of Stuttgart

The model scale wind tunnel of University of Stuttgart is suitable for aerodynamically measure vehicle models in 20% scale and 25% scale. The wind tunnel is equipped with a 5-belt rolling road system featuring a steel center belt and four wheel drive units with polymer belts. Boundary layer suction, starting at the nozzle exit plane, and tangential blowing in front of the center belt are used to minimize the height of the boundary layer [1–3] (Fig. 6).

Fig. 6. Test section of the model scale wind tunnel with the DrivAer model and the FKFS **swing**® system

The wind tunnel can achieve a maximum speed of 80 m/s (288 km/h) with a fan power of 335 kW. The nozzle has a size of 1.6 m × 1.0 m and the test section is 2.5 m long. The wind tunnel is equipped with an underfloor balance, capable of measuring both steady state and transient forces acting on the model. With the FKFS ***swing*** system it is possible to generate a transient flow field in the wind tunnel, either generated by harmonic signals or by white noise with frequencies of up to 10 Hz [4].

For validation, the DrivAer model [5, 6] is measured with the estate rear end and pressures on the side of the rear end as well as in the centerline of the rear end are evaluated. The rear ends of the DrivAer models available at University of Stuttgart can be equipped with 64 pressure tabs that connect to a PSI initium system with the scanners located inside the model. The position inside the model results in a very short tubing between the pressure tab and the scanner, allowing to measure pressures at high frequencies with a high accuracy. This offers an ideal starting point for a validation of the new high frequency pressure strip system. In a first test, the vehicle is measured with full road simulation at 180 km/h under different yaw angles of up to 30°.

4.2 Validation Results

To validate the new high frequency pressure strips, several experiments were conducted, both in a specialized test stand and in the model scale wind tunnel. With the specialized test stand, a direct comparison of different pressure transducers is possible in a well-known and well-controlled environment, where no external influences on the measurement results are expected.

The tests in the model scale wind tunnel on the other side show the capability for pressure measurements in real life. Here, a direct comparison between the different sensor types is not 100% possible, as the pressure strips taped to the outside of the vehicle can influence the flow around the vehicle. The pressure tabs connected to the PSI-system are small holes drilled into the surface. They are connected via tubing to the pressure transducers inside the model. The only influence to the flow around the vehicle is due to the tube for reference pressure and the signal cable, both exiting the vehicle in the wake of the rear wheel. Still, it is important to know if the influence of the new pressure strips to the flow around the vehicle is small enough to still be able to use the pressure strip system, even on small-scale models.

4.2.1 Test Stand Validation

The first validation step was to check the overall accuracy of the pressure strips by comparing the measured pressure amplitude in the sealed volume behind the speaker to the results from the PSI initium system and the Kulite sensors. Two reference systems were used here, to ensure that the reference results are reliable. The experiments were conducted using sinusoidal signals at discrete frequencies from 1 Hz to 20 Hz.

Figure 7 shows the measured pressure amplitudes for a given frequency of 10 Hz. The Kulite and PSI initium sensors - as reference sensors - show a very small standard deviation between the eight different measurement points (0.04 Pa for PSI, 0.2 for Kulite), also the delta between the highest and lowest value is quite small (0.1 Pa for PSI, 0.6 for Kulite). The standard deviation for the sensors of the new pressure strips is higher (0.8 Pa) and also the maximum delta between highest and lowest value (2.6 Pa), still it is much better than the specified measurement accuracy of the pressure strips.

The results are comparable for all tested frequencies. In Fig. 8, the results of the respective eight sensors are averaged to get an overall comparison between PSI, Kulite and the pressure strips. The plot shows the deviation of the measured pressure between

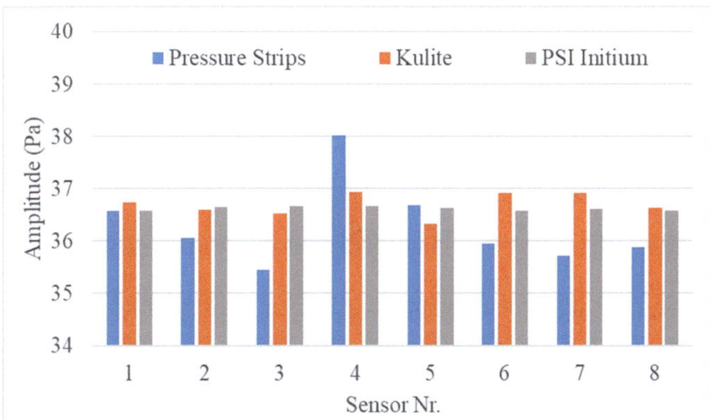

Fig. 7. Measured amplitude for each of the 8 sensors at 10 Hz

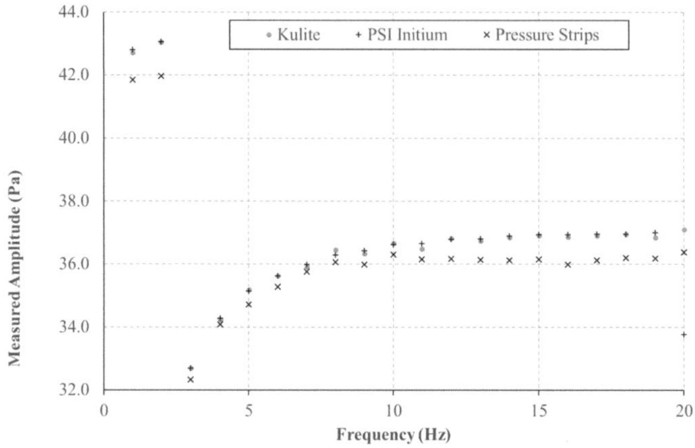

Fig. 8. Comparison of different pressure measurement systems – measured amplitudes at different frequencies

pressure strips and the reference values. With a maximum deviation of around 1.5 Pa, the results again are better than the specification.

These results have to be considered with respect to the purpose of the system: The pressure strip system is designed so that it can be installed very fast and the overall handling of the system is designed to be as easy as possible. As the pressure strips are only taped to the surface, there is no need to modify the measurement object (in most cases the vehicle) in any way for installing the sensors. For the other systems, holes have to be drilled into the model and the sensors have to be connected to the holes via tubing. For this simplicity, some minor disadvantages in overall accuracy can be accepted. Therefore, with the very high flexibility of the pressure strip system, the achieved accuracy is very good.

In the second validation step, the dynamic response of the sensors was tested. Therefore the measured frequencies of the three systems were compared to the setpoint frequency. In terms of frequency response, all of the measurement systems show a comparable and very good performance (refer to Fig. 9). The frequency deviation from the setpoint is the same for all sensors, and therefore the sensors are rather showing the response characteristic of the test stand rather than the measurement uncertainty of one of the systems.

4.2.2 Wind Tunnel Validation

To validate the usage of the pressure strip system in a wind tunnel test, a quarter scale model was equipped with four of the new pressure strips: One on the centerline of the roof, one on each side of the rear end and one below the rear window. The positions are displayed in Fig. 10. The cables of the pressure strips were taped to the surface of the model, then continuing along the test section floor and finally lead to the control unit placed outside the flow.

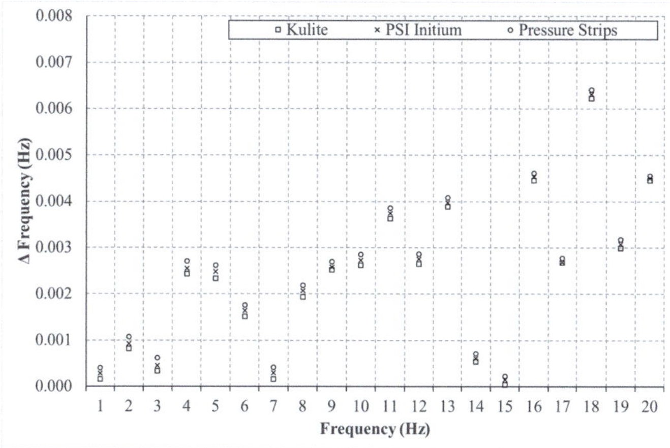

Fig. 9. Measured frequency deviation with the different pressure measurement systems

For the PSI-system 21 pressure tabs are used in total, located in the same position as the pressure strips: seven on the centerline of the roof and seven on each side of the rear end. The pressures on the rear window were not measured with the PSI system and can therefore not be used for validation.

The pressures on the vehicle's surface were measured at different yaw angles between $\beta = -30°$ and $\beta = +30°$. In Fig. 10 the pressures measured at 0° yaw and at ±30° yaw are plotted to show the range of pressure measured on the vehicle. For the

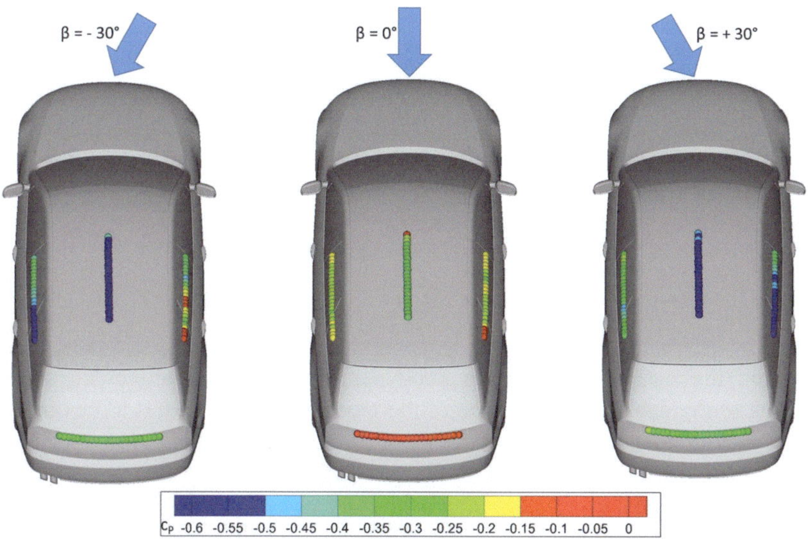

Fig. 10. Surface pressure measured at different yaw angles

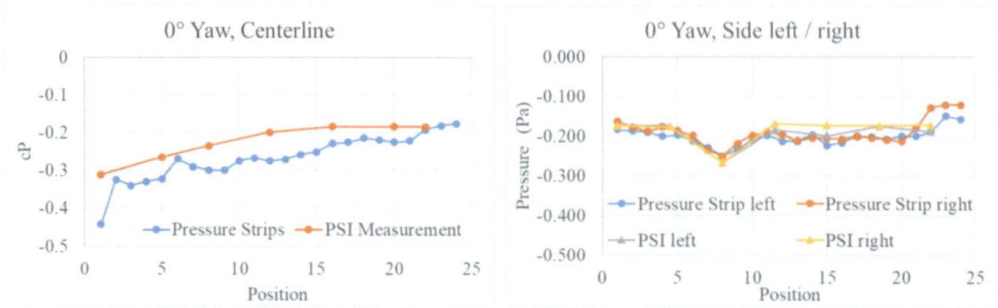

Fig. 11. Surface pressures measured at 0° yaw (position 1 is in the front, 24 in the rear)

direct comparison between pressure strips and PSI, two different yaw angles will be examined in detail in the following paragraphs: $\beta = 0°$ and $\beta = 15°$.

When comparing the results for the $\beta = 0°$ yaw case, some offset can be seen in the centerline (refer to Fig. 11), which can be ascribed to the interference caused by the thickness of the pressure strips. The pressure on the sides of the model are comparable for both measurement systems both in trend as well as in the absolute values. The pressures on both sides of the vehicle are the same, due to the symmetry of the top part of the model.

In the $\beta = 15°$ yaw-case (refer to Fig. 12), the influence of the taping in front of the centerline pressure strip is reduced and both measurement techniques result in the same pressures. Also the pressures measured on the vehicle's sides are comparable for both measurement systems. The pressures on the side are now no longer symmetrical, as the pressure on the leeward side is decreasing while the pressure on the upwind side of the vehicle is increasing. This effect is increased with higher yaw angles.

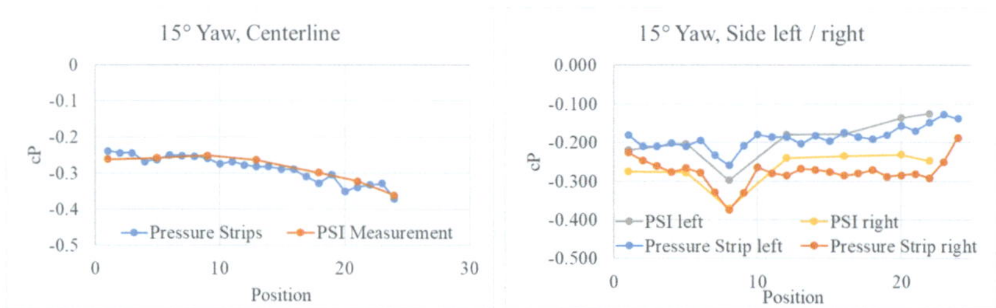

Fig. 12. Surface pressure measured at 15° yaw (position 1 is in the front, 24 in the rear)

Overall, the results show a certain aerodynamic influence of the pressure strips and the taping of the cables to the surface of the model. However, it has to be considered that the measurements were performs in quarter scale, resulting in a much higher influence than in full-scale. Still, the taping of the cables should also be considered in full-scale applications, especially if the cables are placed in an area sensitive to flow separation. Here an intelligent placement of the cable can minimize the influence of the system to the aerodynamics of the vehicle.

5 Conclusions

A cost effective and simple to use pressure measurement system for unsteady flow measurement was successfully developed. Several specifics of vehicle productive testing were aimed such as rapid preparation, instantaneous error reporting, or simple migration between stages of experiment. All stated goals were met resulting in a user-friendly powerful tool for flow investigation. The new system is not competing with current unsteady pressure measurement tools but rather offers an alternative approach of large number of sensors with some accuracy sacrificed.

The system is capable of operation with a 12 VDC supply and is therefore suitable for on-road vehicle testing, although such tests did not take place yet. The general usage of the system reaches far beyond road vehicle aerodynamics and may find its place in other fields of experimental fluid mechanics.

Multiple tests were carried out during the development process with the final validation results shown in this paper. The validation results show that the new pressure strip system is giving good results both in steady state and in transient conditions. The performance of the pressure strips during validation was always better than the specifications. Therefore, with the new pressure strip system a valuable tool is available for very easy and fast surface pressure measurements.

References

1. Filipský, J., Polická, P., Čížek J.: Multipoint pressure strip measurement for advanced CFD validation. In: FKFS Conference "Progress in Vehicle Aerodynamics and Thermal Management", Stuttgart (2015)
2. Potthoff, J.: Die IVK-Kraftfahrzeug-Windkanalanlage der Universität Stuttgart. Haus der Technik-Conference "Aerodynamik des Kraftfahrzeugs", Essen (1987)
3. Wiedemann, J., Potthof, J.: The New 5-Belt Road simulation system of the IVK wind tunels - design and first results. SAE Paper# 2003-01-0429, Detroit (2003)
4. Wittmeier, F.: The recent upgrade of the model scale wind tunnel of University of Stuttgart. SAE Int. J. Passeng, Cars – Mech. Syst. **10**(1) (2017). http://doi.org/10.4271/2017-01-1527
5. Stoll, D., Schoenleber, C., Wittmeier, F., Kuthada, T., et al.: Investigation of aerodynamic drag in turbulent flow conditions. SAE Int. J. Passeng. Cars - Mech. Syst. **9**(2), 733–742 (2016). https://doi.org/10.4271/2016-01-1605
6. Heft, A., Indinger, T., Adams, N.: Introduction of a new realistic generic car model for aerodynamic investigations. SAE technical paper 2012-01-0168 (2012). https://doi.org/10.4271/2012-01-0168
7. Wittmeier, F., Kuthada, T.: Open Grille DrivAer model - first results. SAE Int. J. Passeng. Cars - Mech. Syst. **8**(1), 252–260 (2015). https://doi.org/10.4271/2015-01-1553

Erratum to: The Aerodynamics Development of the New Land Rover Discovery

Sébastien Chaligné, Ross Turner, and Adrian Gaylard

Erratum to:
Chapter "The Aerodynamics Development of the New Land Rover Discovery" in: J. Wiedemann (ed.), *Progress in Vehicle Aerodynamics and Thermal Management*, https://doi.org/10.1007/978-3-319-67822-1_9

In the original version of the book, the title of Chapter 9 has been changed from "The Aerodynamics Development of the New Land Rover Discovery 5" to "The Aerodynamics Development of the New Land Rover Discovery". The erratum chapter and the book have been updated with the change.

The updated online version of this chapter can be found at
https://doi.org/10.1007/978-3-319-67822-1_9

J. Wiedemann (ed.), *Progress in Vehicle Aerodynamics and Thermal Management*,
https://doi.org/10.1007/978-3-319-67822-1_18

Author Index

J. Wiedemann (ed.), *Progress in Vehicle Aerodynamics and Thermal Management*,
https://doi.org/10.1007/978-3-319-67822-1

FSC
www.fsc.org
MIX
Papier aus verantwortungsvollen Quellen
Paper from responsible sources
FSC® C105338

If you have any concerns about our products,
you can contact us on
ProductSafety@springernature.com

In case Publisher is established outside the EU,
the EU authorized representative is:
Springer Nature Customer Service Center GmbH
Europaplatz 3, 69115 Heidelberg, Germany

Printed by Libri Plureos GmbH
in Hamburg, Germany